KGB TRAINING MANUAL

RECRUITING AGENTS

ВЕРБОВКА АГЕНТУРЫ

1969

**Translated from the original Russian
by Major Christoph P. Schwanitz (Ret.)**

**Conflict Research Group
London, 2025**

Original edition published for internal use by the Ministry for State Security (KGB) of the Union of Soviet Socialist Republics.

This English Language translation published by Conflict Research Group, London, United Kingdom, 2025

Copyright Notice

About the Conflict Research Group

Conflict Research Group (CRG) is a non-profit think-tank based in the United Kingdom, dedicated to advancing understanding of the art and science of Unconventional Warfare. With a focus on the academic study of guerrilla warfare, revolutionary warfare, asymmetric warfare, Fourth Generation Warfare, Fifth Generation Warfare and political unrest, CRG's work sheds light on the complexities and nuances of modern conflicts. By bringing critical and key works back into print, the organization serves as a vital resource for academics, policymakers, and military professionals seeking in-depth knowledge in these specialized fields.

At the heart of CRG's mission is the belief that a comprehensive understanding of Unconventional Warfare is essential for addressing contemporary security challenges. The group's research and publications delve into historical and contemporary case studies, exploring the strategies, tactics, and implications of irregular warfare. Through this rigorous analysis, CRG contributes to the development of more effective and adaptable strategies for dealing with non-traditional threats.

One of the key aspects of CRG's work is its publishing arm, which is dedicated to bringing into print seminal works on Unconventional Warfare. The group's publications cover a wide range of topics, from historical accounts of guerrilla movements to theoretical analyses of contemporary conflict dynamics and of course reprints of historical official publications. By making these works accessible to a broader audience, CRG aims to enrich the discourse on Unconventional Warfare and contribute to the development of more nuanced and effective approaches to resolving conflicts and disrupting, degrading and defeating unconventional threats.

CRG's research is categorised by its interdisciplinary approach, drawing on insights from military history, political science, sociology, and international relations. This holistic perspective allows the organization to address the multifaceted nature of unconventional warfare, considering not only military tactics, but also the granularity of the political, social, and economic dimensions of conflicts. Through this comprehensive approach, CRG provides a deeper understanding of the root causes and long-term implications of irregular warfare.

Publisher's Note

This English translation of *Recruiting Agents* (1969) is based on a Soviet-era KGB manual that was never declassified. The original document, produced by the KGB's First Chief Directorate, was intended for internal use in intelligence training and operational instruction. While the Soviet Union no longer exists, this text serves as an important historical resource on Cold War espionage tradecraft.

The translation presented here is the result of meticulous effort to ensure accuracy, clarity, and readability for an English-speaking audience. It includes refinements, contextual clarity, and structural modifications to make the material accessible without altering the meaning of the original text. Additionally, this edition features a *Translator's Note*, *Editor's Introduction* and an explanatory chapter about the KGB to provide historical and operational context.

Neither the translator or the publisher endorse or promote espionage activities but present this work purely for educational, historical, and academic research purposes.

Translator's Note

I grew up in East Germany during the 1970s and 80s at the height of the Cold War, with an uncle who was a high-ranking official in the notorious and now thankfully defunct East German Ministerium für Staatssicherheit or Ministry for State Security, more commonly known today as the Stasi.

I did not know it at the time, but before moving into his highly prestigious and important role as Stasi chief for East Berlin, my uncle had spent a large part of his career in the Stasi's Hauptverwaltung Aufklärung section (HVA) which dealt with foreign intelligence operations. This was the GDR's equivalent of the Soviet KGB or the American CIA or the British MI6.

As a young man, particularly during that chaotic period immediately following German Reunification, this all conspired to give me a deep interest in the intricacies of intelligence operations and ensured that I would later seek to forge a career in the field of intelligence. Of course, I would go on to do just that, serving with several different operational units within the German Federal Army until my retirement from the Bundeswehr in 2017.

As someone who had spent decades studying Russian language, culture, military and intelligence structures in the course of my work as an officer in the Army, I knew that translating these manuals would require more than just basic linguistic proficiency. It would demand an intimate knowledge of the nuances of each language, as well as a deep understanding of the cultural, historical, technical and operational contexts in which they were written.

One of the more significant challenges I faced, and the one which is least likely to be of any great interest to a reader of this book, was navigating the complexities of Russian grammar and syntax. Unlike German, which is known for its strict rules and conventions, Russian has a more relaxed system that can make it difficult at times to accurately convey meaning. For example, Russian word order often prioritizes grammatical function over semantic content, making it essential to carefully consider the context in which each sentence appears.

Furthermore, Russian relies heavily on prepositions and case endings to convey subtle shades of meaning, whereas German tends to rely more on verb conjugation and adverbial phrases. This meant that I had to be particularly mindful when translating individual words or phrases, as their meanings could shift significantly depending on the surrounding context.

Another significant challenge was capturing some of the more obscure technical jargon used in these manuals. With many of the KGB manuals in the cache dating to the 1960s and 70s, some of the old Soviet terminology has become obsolete and has been replaced by other terms within the Russian Federation intelligence services. The manuals display a wide array of obsolete and specialized terms for various aspects of intelligence operations, from types of agents to counter surveillance techniques to clandestine communications methods.

As someone who is very familiar with German and other NATO partners' intelligence operations and with their jargon and acronyms, I found myself constantly referencing my own knowledge base to ensure that I accurately conveyed the intended meaning of any given passage that included technical operational language. In those few cases where I could not be 100% sure of a technical term's meaning, I simply extrapolated to the best of my ability.

The fact that many of these manuals remain classified in Russia even today speaks volumes about their significance and relevance. It's likely that some are still being used by Russian Federation intelligence services to train new personnel, while others would have no doubt been declared obsolete but remain sensitive due to the nature of their contents.

I have a responsibility to ensure that these manuals are translated accurately and with appropriate sensitivity. It's not just about conveying technical information; it's also about respecting the cultural and the operational context in which they were written. In many ways, translating these KGB tradecraft training manuals was akin to conducting an archaeological excavation into the past. Each sentence or phrase revealed a piece of history that had been hidden away for decades, waiting to be uncovered and shared with the world.

As someone who has spent years studying Russian language and culture as well as evaluating the potential threats which an adversarial Russian Federation might in the future pose to my homeland and our NATO partners, I'm proud to have played a role in making this significant historical material available for public consumption.

I would like to thank "DC" and "CB" from Conflict Research Group for assigning me the delicate but critical task of translating this important material. Having become well informed of the vital work being undertaken by Conflict Research Group, I am honoured to be of service even in this small way.

I would like to thank my beloved wife, Birgitt, for dealing with my many absences and long days spent locked away in my study working on this material and accepting it all with grace and good humour.

I would like to also thank Birgitt for her assistance in helping me translate certain more complex passages from German to English and for proof-reading the final manuscript to correct my abysmal English language grammar. As always, without her, I would be diminished.

Please note that any errors or omissions in these translated pages which may serve to detract from the original Russian language documents are mine and mine alone.

Christoph P. Schwanitz,
Major, KSA (ret.)
Görlitz, 2025

About the KGB

The KGB was the foreign intelligence and domestic security agency of the Soviet Union. It was established on the 13th of March, 1954, soon after the death of Soviet dictator Josef Stalin and it was dissolved with the fall of the Soviet Union on the 3rd of December 1991. The KGB's First Main Directorate was split off and became the Russian Federation's current foreign intelligence service, the FSB.

In addition to its primary responsibilities for foreign intelligence and domestic counterintelligence, during the Soviet era the KGB also had duties such as safeguarding the country's political leadership, overseeing border troops, and carrying out surveillance of the population.

In this book, we are dealing solely with the foreign intelligence aspects of KGB operations, so we shall look at the KGB's foreign intelligence apparatus.

The KGB's First Main Directorate, also known as the First Chief Directorate, was responsible for intelligence operations outside of the Soviet Union.

The directorate was organised into various directorates, including:

Directorate "R" -   Planning and Analysis,
Directorate "S" -   Illegals,
Directorate"T" -   Scientific and Technical Intelligence,
Directorate "K" -   Counter-Intelligence,

Directorate "OT" -   Operational and Technical Services,
Directorate "I" -   Computers Service (would
                          be known as "IT" today),
Directorate "A" -   Active Measures,
Directorate "RT" -   Operations within the USSR

In addition to the administrative directorates listed above, the First Main Directorate had various "Desks" or "Departments" dedicated to operations in various parts of the world or other specialised functions. These were:

1st Department -   North America
2nd Department -   Latin America
3rd Department -   UK, Australia, NZ, Scandinavia,
                    Malta
4th Department -   East Germany, Austria, West
                    Germany
5th Department -   France, Spain, Portugal,
                    Luxembourg, Switzerland, Greece,
                    Italy, Yugoslavia, Albania, Romania
6th Department -   China, Laos, Viet Nam, Cambodia,
                    North Korea, South Korea
7th Department -   Thailand, Indonesia, Malaysia,
                    Singapore, Japan, Philippines
8th Department -   Afghanistan, Turkey, Iran, Israel
9th Department -   English-speaking countries in
                    Africa (South Africa, Rhodesia/
                    Zimbabwe, Tanzania, Nigeria, etc.)
10th Department -   French-Speaking Countries in Africa
11th Department -   Liaison with other communist
                    countries' intelligence services
                    particularly Cuban and Warsaw Pact
                    nations (was previously known as
                    the "Advisor's Department")
12th Department -   Covers
13th Department -   Covert Communications
14th Department -   Forgeries
15th Department -   Operational files and archives

16th Department -	Signals intelligence
17th Department -	India, Pakistan, Bangladesh, Sri Lanka, Burma, Nepal
18th Department -	Egypt, Syria, Libya, Iraq, Oman, Saudi Arabia, Kuwait, Sudan, Jordan, Morocco, United Arab Emirates/Trucial States
19th Department -	Soviet Expatriates and Emigres
20th Department -	Liaison with 3rd World / newly independent states

It is the KGB's First Main Directorate which was the publisher of these manuals and is most likely that they were produced by staff of the First Main Directorate's *Directorate OT*, which was responsible for Operational and Technical Support functions.

KGB foreign intelligence networks were operated by a KGB Residence or *Rezidentura* as it is known in phonetic Russian. Please note that in these translations, we sometimes refer to the Residencies using the equivalent CIA term "Station". This is simply to reduce the possibility of confusion and to differentiate between a Residence and a private residence such as those used as safe houses or clandestine postal addresses. Similarly, within the translation in these situations, we will refer to the KGB Residence's "Resident" using the CIA term "Station Chief".

The KGB Resident or Station Chief was a legal intelligence officer usually operating under diplomatic cover as a "cultural attache" or similar. Diplomatic credentials gave the Resident diplomatic immunity meaning the security forces of the country in which he was operating could never arrest a KGB Resident. At best they could have him expelled from the country like any other diplomat, but this usually had serious diplomatic consequences. Instead, most countries usually worked out fairly quickly who was KGB within their local Soviet embassy and they usually allowed the KGB Resident to operate, but placed him and other Soviet embassy staff under heavy counterintelligence surveillance.

Typically, a KGB Residency was organised into different sections or "lines". Each section had a separate function which supported operations conducted out of any given Residency. These sections could be further categorised into separate functions - Operational and Support.

Operational sections of a KGB Residency were as follows:

Section "EM" -	Intelligence and surveillance of the activities of Soviet Emigres in the host country
Section "KR" -	Counterintelligence and protective security of the Residency
Section "N" -	Support to "illegal" Intelligence Officers in the host country
Section "PR" -	Economic, military, political intelligence on the host country or region as well as active measures such as black propaganda
Section "SK" -	Surveillance and reporting on Soviet diplomatic staff in the host nation.
Section "X" -	Technical intelligence and advanced technology acquisition and transfer.

Support sections of a KGB Residency were as follows:

Section "OT" -	Technical support
Section "RP" -	Signals intelligence
Section "I" -	Information technology

Support staff not assigned to their own specific section included drivers, signals operators, cipher clerks, administrative staff, finance personnel.

Table of Contents

PART I:
AGENT RECRUITMENT AS A CORE FUNCTION OF INTELLIGENCE OPERATIONS

PART II:
IDENTIFYING AND SELECTING TARGETS FOR RECRUITMENT

PART III:
THE RECRUITMENT PROCESS AND ITS STRATEGIC COMPONENTS

PART IV:
DEVELOPMENT, EXECUTION, AND FINALIZATION OF RECRUITMENT OPERATIONS

Editor's Introduction

The original Russian language manual this English translation is based on was found on the deep web in a cache of scanned older Soviet KGB training materials in a folder on a Russian language .onion site. It is believed that these materials were posted by a dissident many years ago, perhaps even as long ago as 2010 or 2012 based on file metadata. The cache was later posted on the surface web, where to this day scans of the original Russian language documents can still be found through a simple search on any search engine.

Various think-tanks from English-speaking countries had made promises to translate and publish these materials, but despite waiting over five years for them to do so, no apparent progress has been made. With the Russian invasion of Ukraine in February 2022, it appears that translation and publication of the KGB training manuals is no longer a priority for these organizations. As a consequence, and with no clear end to the Ukraine War in sight at the time of writing, we have gone ahead and translated and published the KGB manuals from the cache ourselves.

Please note that we are not the first to publish English language translations of some of these materials. Circa 2020, enterprising persons unknown, in a blatant cash-grab, ran a couple of these documents through some translation software, probably Google, before dumping the resulting unedited text into a book format for publishing on Amazon. We purchased a copy of each of these translations to see whether there would still be a requirement for our professionally translated editions. Sadly, all were largely unreadable, therefore, we pushed ahead with our project.

As Conflict Research Group deals mostly with unconventional warfare, resistance, and inform/influence operations from the perspective of non-state actors, it would seem to the casual reader that espionage training materials from a former nation-state intelligence agency such as the Soviet KGB would fall well outside our remit.

This is simply not the case. During the Cold War, the Soviet Union, its Warsaw Pact satellites and other communist states such as the People's Republic of China and the Democratic People's Republic of Korea invested many billions of dollars in to supporting subversive and revolutionary groups fighting against western interests from Southeast Asia to the Middle east, to Latin America, to Southern Africa. Soviet support for such groups was not limited to weaponry and war materiel, but also included training in communist political theory, revolutionary and guerrilla warfare and of course, in clandestine tradecraft to allow members of a revolutionary or terrorist group to organize, plan and conduct their activities in secret.

Western-trained security forces typically used extremely effective British, French or American counter-intelligence and counter-insurgency methods to detect and destroy insurgent undergrounds or espionage rings at or before their nascent stage, so there was a requirement for guerrilla or terrorist groups sponsored by the Soviet Union to be given the most effective tradecraft training available in the communist world, and that came from the KGB's First Main Directorate.

Two English language resources which closely follow KGB procedures and concepts can be found in the 1980s-era South African Communist Party pamphlet *How to Master Secret Work* and in the 1970s-era document *Security and the Cadre* produced by a Puerto Rican separatist group operating in the US, the Fuerzas Armadas de Liberacion Nacional (FALN). Anyone reading through those two sources and then reading one of these KGB manuals will soon find examples which appear in all three, sometimes almost word-for-word.

Unlike some Western intelligence services such as the CIA, which train personnel in very specific, complex tradecraft techniques and methodology (some involving literal magician's sleight of hand), the KGB instead concentrated on teaching its personnel general concepts. This forced the KGB operative to become highly adaptable and imaginative in putting those concepts into action in the field. This lack of a "toolkit" of relied-upon tactics, techniques and procedures meant no clear patterns were set, making it just that much harder for western counterintelligence services to anticipate the specificities of a KGB intelligence officer or agent's tradecraft in the field.

In closing, I would like to thank Major Chris Schwanitz for his accurate translations of these materials, as well as for standing by for almost 18 months while we decided whether or not to go ahead with this project. I would also like to thank "DC" and her OSINT team for backtracking the circumstances of how the original scanned documents came to be posted online. Finally, I would like to thank you, the reader, for your interest in this "charming vintage spycraft" and for your support of CRS by purchasing this book.

CB
London, 2025

INTRODUCTION

The effectiveness of the intelligence services of socialist states in achieving their strategic objectives hinges primarily on the proper organization of operations across all areas of activity within capitalist nations. Success is determined by the caliber of intelligence officers and agents, as well as the ability of KGB personnel to identify, cultivate, and recruit suitable individuals for clandestine work, subsequently integrating them into the agent network.

The agent relationship constitutes a structured, covert arrangement between an intelligence officer and an individual enlisted for intelligence purposes. Upon consenting to clandestine collaboration, this individual consciously undertakes tasks assigned by the intelligence service, thus formally assuming the role of an agent.

Once recruited, the agent undergoes continuous ideological and political indoctrination to reinforce commitment to the Soviet cause. The intelligence service methodically expands the agent's operational capabilities, instructing them in the art of tradecraft, including secure communication, counter-surveillance, and the execution of intelligence tasks in a clandestine manner. The agent is also trained to maintain proper conduct within professional and social circles, ensuring seamless integration into daily life while avoiding suspicion.

A well-prepared agent fully understands the gravity of their covert affiliation and the necessity of maintaining absolute secrecy. Through adherence to rigorous operational security measures, the agent ensures the successful execution of assignments while mitigating exposure.

By instilling unwavering ideological loyalty, refining operational skills, and enforcing strict adherence to tradecraft principles, the intelligence service maximizes the agent's effectiveness while minimizing operational risks. In this way, the agent becomes a dedicated and disciplined instrument of Soviet intelligence, fulfilling assigned tasks with precision and discretion in the ongoing struggle against imperialist adversaries.

In the course of intelligence work, agents may be rendered inoperative for various reasons—some due to natural attrition, while others are removed from the agent network due to operational, security, or administrative concerns. As a result, the intelligence service must continuously recruit new agents to maintain and expand its network, ensuring the uninterrupted execution of its objectives.

The act of securing an individual's consent for clandestine cooperation—i.e., recruitment—is merely the final phase of an extensive preparatory process. This process involves a series of calculated steps: conducting reconnaissance on a target of interest to identify individuals within its orbit who either possess or could potentially acquire intelligence value; assessing the most viable recruitment candidates; evaluating their motivations and vulnerabilities to determine the most effective approach; and selecting the optimal means of introducing a recruiter into their sphere of influence.

All preparatory activities culminate in the recruitment conversation, during which the intelligence officer secures the individual's agreement to work covertly for the service. This moment marks the transition from cultivation to formalized agent relations. Thus, recruitment itself constitutes the core of recruitment operations and is the most decisive phase of agent-network expansion.

The acquisition of agents is one of the most critical and high-stakes components of intelligence work. Without a robust and reliable agent network, intelligence collection cannot be effectively conducted. Likewise, without sustained, methodical, and strategic recruitment efforts, it is impossible to develop an agent network capable of fulfilling intelligence tasks with efficiency and security.

However, poorly executed recruitment can have severe consequences. A mismanaged approach or failure to successfully enlist a citizen of a capitalist state may not only compromise the recruiter's cover but could also lead to diplomatic complications between the socialist country and the target state. Such failures risk exposing intelligence operations, damaging broader intelligence objectives, and forcing operational standdowns, thereby inflicting long-term harm on the intelligence service's ability to function effectively.

Recruiting citizens of capitalist states to cooperate with a socialist intelligence service is inherently challenging. Most individuals are reluctant to engage in clandestine work for a foreign intelligence agency, given the associated risks and the potential consequences of exposure. To secure their cooperation, intelligence officers must exert influence over the recruit's worldview and psychological disposition, gradually steering them toward a favorable decision. Persuasion, pressure, and tailored psychological approaches are all tools at the disposal of a skilled recruiter.

A prospective recruit must come to recognize that clandestine cooperation entails personal risk and demands strict adherence to discipline. Participation in intelligence work requires unwavering commitment and the ability to operate under conditions of secrecy and control. For this reason, agent recruitment can only be conducted by intelligence officers who possess advanced political training, operational expertise, and the necessary personal qualities to assess, influence, and successfully recruit individuals for covert work.

Like all agent-operational activities, recruitment efforts require a thorough understanding of the operational environment in the target country. Intelligence officers must take into account local conditions, navigate legal and social constraints, and neutralize obstacles posed by hostile intelligence and counterintelligence agencies. A skilled recruiter must have a clear grasp of their objectives, possess the ability to assess and analyze potential candidates, and determine the most effective means of securing their cooperation. When the opportunity for recruitment arises, the intelligence officer must execute the approach with precision, employing the appropriate tradecraft to establish agent relations while maintaining operational security.

Within intelligence operations conducted by socialist states, two primary types of relationships are cultivated with citizens of capitalist countries for intelligence purposes: agent relationships and trusted contacts. Only the former—fully recruited, trained, and directed agents—provide intelligence services with the degree of control, reliability, and access necessary to achieve strategic objectives. While trusted contacts may provide useful insights or indirect support, they lack the operational discipline and structured guidance of a true agent. Furthermore, reliance on untrained or informally controlled individuals for sensitive intelligence work increases risk and reduces the effectiveness of covert operations.

Thus, to ensure the security, success, and continuity of intelligence work, recruitment efforts must prioritize the identification, cultivation, and formal recruitment of fully committed agents. Trust-based relationships, while occasionally useful, cannot substitute for a well-developed agent network operating under strict clandestine discipline.

PART I

AGENT RECRUITMENT AS A CORE FUNCTION OF INTELLIGENCE OPERATIONS

Objectives of Recruitment Operations

The primary mission of the intelligence services of socialist states is to identify and neutralize hostile actions planned by adversary powers, disrupt subversive activities directed against socialist nations, and actively shape political, economic, and social developments within capitalist states in a manner that advances socialist interests.

To successfully achieve these strategic objectives, it is imperative to establish and maintain a robust network of highly reliable and ideologically committed agents. The effectiveness of an intelligence service is directly proportional to the quality and breadth of its agent network. Therefore, the intelligence services of socialist states devote considerable resources to both expanding the numerical strength of their agent network and enhancing the operational capabilities of recruited agents. The size and structure of the agent network are determined by the scope of intelligence operations and the strategic priorities dictated by the state.

Recruitment is the cornerstone of intelligence operations. Without a continuous and systematic effort to identify, cultivate, and recruit new assets, it is impossible to sustain the intelligence capacity necessary for long-term success. As such, recruitment is not merely a supporting function but a principal operational objective of intelligence services.

Intelligence officers operating within capitalist states in the domain of political intelligence must prioritize the recruitment of agents positioned to provide access to classified or otherwise restricted information regarding the foreign and domestic policies of their host government. These agents serve as critical sources for obtaining insights into decision-making processes, diplomatic strategies, and internal political dynamics.

The recruitment of well-placed individuals within political institutions, diplomatic circles, and policymaking bodies allows socialist intelligence to anticipate adversarial actions, influence political discourse, and, when necessary, implement measures to counteract initiatives that run contrary to socialist interests. Thus, political intelligence officers must be particularly adept at identifying, assessing, and recruiting individuals who can provide actionable intelligence and facilitate broader operational goals.

By systematically strengthening its agent network through targeted recruitment efforts, the intelligence service ensures its continued ability to conduct effective intelligence collection, counter hostile activities, and exert strategic influence over geopolitical developments in capitalist states.

Intelligence officers of socialist states operating in capitalist countries within the sphere of political intelligence must prioritize the recruitment of agents who can provide access to classified information regarding the foreign and domestic policies of the target government. These agents serve as vital sources for acquiring intelligence on strategic decision-making, policy formulation, and geopolitical maneuvering.

To achieve this, intelligence officers must establish agent networks within key institutions where government policies are formulated and where sensitive political documents are concentrated. The highest-priority recruitment targets include personnel in the cabinet of ministers, the Ministry of Foreign Affairs, leadership structures of political parties, and executive offices of major monopolistic corporations.

Within the Ministry of Foreign Affairs, efforts should focus on recruiting agents positioned in critical areas such as the ministerial secretariat, offices of deputy ministers, the political affairs department, and particularly divisions that handle socialist bloc relations. Additionally, agents embedded in the cipher department and the ministry's classified archives provide unparalleled access to secret diplomatic communications and sensitive documents of direct operational interest.

Intelligence officers engaged in scientific and technical

intelligence (S&T) must adopt a similarly targeted approach. For example, in executing directives to obtain classified information on thermonuclear warheads for intercontinental ballistic missiles (ICBMs), the recruitment strategy should focus on individuals positioned within the design bureau, plant management, and production workshops responsible for manufacturing next-generation weapons systems. Personnel within these key sectors possess privileged access to technical specifications, research data, and classified production details, making them prime targets for recruitment.

By securing agents in these critical nodes, intelligence officers ensure a steady flow of high-value technical intelligence essential for advancing the strategic objectives of the socialist bloc.

One of the most crucial yet complex challenges in intelligence operations is the infiltration of enemy intelligence and counterintelligence organizations. Effective counterintelligence efforts against adversaries of the socialist bloc depend on deep penetration into the adversary's own agent networks and operational structures.

Securing agents within the intelligence and counterintelligence apparatus of the enemy is among the highest-priority recruitment objectives. However, it is also one of the most difficult, as adversary services actively recruit from the most ideologically hardened and reactionary elements within their societies. Identifying individuals within these structures who are susceptible to recruitment, conducting their assessment, and formulating effective approaches to win their cooperation requires a high degree of skill, patience, and operational finesse.

Despite the challenges, successful recruitment of enemy intelligence and counterintelligence personnel provides unparalleled access to adversary planning, operational methodologies, and agent networks. The intelligence services of socialist states must therefore dedicate significant resources to penetrating these institutions, ensuring that our counterintelligence efforts remain proactive rather than reactive.

The intelligence services of socialist states must prioritize the recruitment of agents within the leadership structures of hostile émigré organizations and opposition groups operating abroad. These organizations, known to maintain direct links with and function under the influence of Western intelligence services, pose a persistent threat to socialist interests. By securing agents within their governing bodies, intelligence officers can disrupt adversarial activities, manipulate internal conflicts, and counteract subversive operations targeting socialist nations.

Another critical recruitment objective is the strategic placement of agents in political and economic sectors of capitalist states, enabling direct influence over key aspects of policy and decision-making. To achieve this, intelligence officers must focus on recruiting high-value agents within political leadership, influential business circles, and financial institutions. These agents must possess the authority, networks, and capabilities necessary to execute political maneuvers and economic strategies that align with socialist interests. Through such penetrations, intelligence services can exert covert influence over governmental and corporate decision-making processes, thereby shaping policies and economic outcomes to benefit the socialist bloc.

The primary mission of intelligence officers operating in capitalist states, as well as those within the central intelligence apparatus, is the recruitment of primary agents—key individuals capable of executing high-priority intelligence objectives across all operational domains, including political, economic, military, technical, counterintelligence, and scientific intelligence. These primary agents serve as the foundation of an effective intelligence network, providing access to classified information, enabling strategic influence, and facilitating operational success.

Additionally, intelligence officers must recruit support agents, whose role is to provide logistical, operational, and security support to intelligence operations. These individuals are essential for maintaining secure communications, facilitating clandestine meetings, executing surveillance operations, and performing other vital organizational functions. Without a well-structured support network, the effectiveness and security of intelligence operations would be severely compromised.

The overarching objectives of socialist intelligence services align with the broader geopolitical strategy of the socialist community. Consequently, intelligence officers must conduct systematic, sustained, and highly targeted recruitment efforts within capitalist states to ensure deep penetration into all sectors of interest. This continuous expansion of the agent network is essential for the successful execution of intelligence missions, including espionage, counterintelligence, strategic influence operations, and active measures against adversary governments.

By embedding trusted agents at all levels of enemy infrastructure, socialist intelligence services secure a decisive advantage in the ongoing struggle against imperialist powers. It is through these deeply embedded networks that intelligence officers can preemptively neutralize threats, shape global political dynamics, and safeguard the security and strategic interests of socialist states.

Identification of the Recruitment Pool

As global society evolves, two distinct historical trajectories have become increasingly evident.

The first trajectory, characteristic of the socialist system, is marked by rapid advancements in productive forces, sustained economic growth, rising living standards, the expansion of democratic principles, and successes in the global struggle for peace. The growing power and influence of socialist states continue to strengthen their position on the world stage, solidifying their ideological and geopolitical authority.

The second trajectory, defining the current state of capitalism, is one of political reaction, economic instability, and growing insecurity among the working class. This system perpetuates social inequality, provokes military conflicts, engages in subversive activities against peace, and exacerbates internal and international contradictions. Economic crises, structural inefficiencies, and social unrest expose capitalism as a system in decline—one that fosters chronic unemployment, mass impoverishment, reckless exploitation of resources, and an ongoing threat of war.

As a result, the increasing political influence of socialist countries generates widespread sympathy and ideological alignment with the socialist cause. This sentiment extends beyond the working class to encompass individuals from various social strata within non-socialist states, forming an extensive recruitment pool for intelligence operations.

The recruitment contingent consists of the social and professional environment from which prospective agents may be selected for covert cooperation with socialist intelligence services. Within capitalist states, this contingent includes individuals who, driven by progressive convictions, actively resist capitalist exploitation through open economic and political struggle. However, it also includes those who, despite harboring anti-imperialist sentiments, refrain from overt opposition due to professional constraints, particularly those employed in government institutions, strategic industries, and military sectors.

Hundreds of millions of people across the world recognize capitalism as an obsolete system characterized by economic chaos, periodic crises, entrenched poverty, and systemic inefficiencies. The imperialist model of governance fuels war, fosters social decay, and has entered an irreversible phase of decline. Within capitalist states, growing dissatisfaction with the ruling class is evident in government institutions, military establishments, research centers, and industrial enterprises, where an increasing number of individuals oppose their own governments and ruling parties, as well as the broader political and economic system in which they operate.

This reality presents a strategic opportunity for socialist intelligence services. The presence of disillusioned individuals within the state apparatus and other key sectors in capitalist countries necessitates a targeted and systematic recruitment effort. Intelligence officers must actively identify, assess, and cultivate these individuals, establishing relationships that will facilitate their eventual recruitment into intelligence work.

By leveraging ideological motivations, professional grievances, and systemic discontent, socialist intelligence officers can integrate such individuals into operational networks, securing invaluable sources of intelligence and furthering the strategic objectives of the socialist bloc.

In the modern era, the struggle between the socialist bloc and imperialist powers has centered on preventing the outbreak of a new war, making this contest the defining feature of contemporary global politics. Across capitalist states, the masses are increasingly engaged in this struggle, aligning themselves with peace movements and opposition efforts against the militaristic ambitions of imperialist governments.

People from all social classes and professional backgrounds are participating in these movements. The ranks of anti-war activists include scientists, engineers, doctors, businesspeople, government employees, industrial workers, religious leaders, and even members of both the petty and large bourgeoisie. Many of these individuals, despite their varying ideological backgrounds, recognize the dangers posed by imperialist aggression and have become active in organizations, political parties, and trade unions advocating for peace.

Socialist intelligence officers must recognize that, with the proper approach, certain individuals within the peace movement can be recruited for intelligence work. However, it is essential to carefully evaluate the intelligence potential of each candidate. In some cases, an individual's public activism, influence, and open denunciation of capitalist militarism may prove more valuable than their covert cooperation as an intelligence asset. Each case must be assessed strategically, ensuring that the most effective means of utilizing a prospective recruit's capabilities are selected.

The imperialist policies of major capitalist states not only harm the working class but also disrupt the interests of certain elements within the bourgeoisie and government bureaucracy. The monopolistic elites of leading capitalist nations impose economic and political pressure on smaller states and subordinate capitalist factions, creating dissatisfaction among sections of the bourgeois class that are negatively impacted by these policies.

Under such conditions, there exists a strategic opportunity to cultivate relationships with disaffected elements within the ruling elite. Some segments of the business and administrative classes, pressured by monopolistic forces, may come to sympathize—albeit indirectly—with the anti-imperialist struggle led by the Soviet Union and its socialist allies.

By skillfully leveraging these contradictions, intelligence officers can, under the right circumstances, recruit or influence individuals from within these circles, securing their cooperation for intelligence or strategic influence operations. These individuals, though part of the capitalist establishment, may be motivated by self-preservation, economic grievances, or opposition to imperialist expansion, making them viable candidates for engagement in intelligence work.

The national liberation struggles unfolding across Asia, Africa, and Latin America have drawn hundreds of millions of people into active political engagement. These revolutionary movements, resisting imperialist domination, naturally gravitate toward the socialist bloc, recognizing it as the principal force providing material aid, strategic support, and moral backing to their cause. While many involved in these struggles may not fully embrace the socialist system, their opposition to Western imperialist powers presents a valuable opportunity for intelligence services to identify and cultivate operatives whose interests align— at least temporarily—with those of the socialist states.

Another significant recruitment source within capitalist states consists of lower-level government employees who handle classified materials yet hail from working-class backgrounds. This includes individuals in roles such as: Cipher clerks, Secretaries, Stenographers, Couriers, Typists, etc. Many of these individuals are democratically inclined and sympathetic to the socialist cause due to their social and economic status. As such, socialist intelligence officers must identify and cultivate these individuals for recruitment, as their access to sensitive government communications and classified information makes them ideal candidates for covert cooperation.

In countries such as the United States, Canada, and throughout Latin America, there are large communities of emigrants from socialist states. Many of these individuals, despite their relocation, have retained deep emotional ties to their homeland and still maintain contact with relatives in socialist countries. Such personal connections create leverage points that socialist intelligence officers can exploit to assess, develop, and ultimately recruit individuals whose sympathies or vulnerabilities make them receptive to cooperation.

By systematically engaging dissatisfied elements within national liberation movements, strategically positioned government employees, and emigrant communities, socialist intelligence services can broaden their recruitment base, ensuring deeper penetration into enemy states and advancing the broader objectives of the socialist bloc.

In capitalist countries, a significant portion of the population, despite being employed, still faces financial hardship. Economic instability, low wages, and rising living costs create an environment in which individuals may be willing to cooperate with the intelligence services of socialist states in exchange for material assistance. Among this economically vulnerable class, there are also those who already sympathize with socialist ideals, making their recruitment even more feasible.

Beyond those motivated by financial need, compromising material can serve as a powerful tool in securing cooperation. Individuals with known corruption, moral failings, or personal misconduct may be leveraged for intelligence purposes through blackmail (Kompromat). Upon receiving credible information about their illicit activities, ethical transgressions, or personal scandals, intelligence officers can offer protection from exposure in exchange for covert collaboration. This method, while effective, requires careful handling to ensure long-term reliability and control over the recruited asset.

From the above, it is evident that capitalist states present a wide and diverse recruitment pool, encompassing individuals from all sectors of society who can be persuaded, influenced, or compelled into clandestine cooperation. However, the size and availability of the recruitment contingent fluctuates over time, influenced by historical, political, and socio-economic factors.

The expansion of recruitment opportunities is largely driven by positive developments within the socialist bloc. The growing power and influence of socialist states, their economic and technological achievements, and their successful involvement in global affairs strengthen their appeal, increasing ideological support among various groups within capitalist societies. These factors broaden the recruitment base, making it easier to identify and cultivate new assets.

Conversely, several negative factors can shrink the recruitment contingent, making intelligence work more difficult. These include:

- Failures and internal crises within socialist states or the global communist movement.
- Aggressive anti-socialist propaganda disseminated by capitalist governments and media.
- Suppression of progressive movements through police crackdowns, government repression, or reactionary political shifts.
- Operational failures by socialist intelligence services, leading to compromised networks and increased counterintelligence efforts by adversary agencies.
- Spy hysteria and enemy exploitation of intelligence failures, fueling widespread paranoia and increasing the difficulty of agent operations.

Given these variables, intelligence officers must not passively rely on the long-term ideological appeal of socialism to sustain recruitment. Instead, they must capitalize on favorable conditions and political shifts—the "tides" of history—that create temporary windows of opportunity for recruitment. Timing is crucial; intelligence services must act swiftly and decisively when circumstances align in their favor, ensuring the continuous replenishment and strengthening of their agent networks.

While intensified police repression, counterintelligence crackdowns, and other adversarial measures may reduce the available recruitment pool and limit opportunities for agent acquisition, they do not eliminate recruitment efforts altogether. These conditions instill fear in weaker individuals, discouraging them from cooperation. However, they also provoke outrage among those with strong ideological convictions, further radicalizing individuals who oppose the imperialist order and motivating them toward active resistance.

In such hostile environments, socialist intelligence officers must adjust their methods, exercising greater discretion and employing enhanced tradecraft to mitigate operational risks. This includes the use of stricter compartmentalization, deeper clandestine methods, and increased reliance on indirect communication to continue recruitment operations without exposure.

Key Factors Affecting the Organization and Execution of Recruitment Operations

The mere existence of a broad recruitment pool within capitalist societies does not, in itself, ensure successful agent acquisition. It is one thing to identify a recruiting base, but another to isolate individuals of operational interest, assess their motivations, and bring them under intelligence control.

The expansion or contraction of the recruitment pool, along with the feasibility of carrying out recruitment operations, is directly influenced by both the domestic intelligence and operational climate of the target country and broader international developments.

In recent years, many capitalist states have significantly tightened security measures:

- Expansion of intelligence and counterintelligence capabilities, leading to greater operational risks.
- Increased state repression against communist parties and progressive organizations, limiting their ability to openly function.
- Suppression of the peace movement and broader anti-imperialist activism, making recruitment among ideological sympathizers more challenging.
- Criminalization of contact between local citizens

and representatives of socialist states, categorizing such interactions as subversive acts.

Nowhere is this repressive environment more evident than in the United States, where a special regime of surveillance and restrictions has been imposed on diplomatic, military, trade, and other official representatives of socialist countries. Their movements are tightly controlled, and they operate under constant surveillance. Furthermore, U.S. citizens suspected of maintaining connections with Soviet representatives or those of other socialist states risk arrest and imprisonment, creating a climate of fear designed to deter recruitment efforts.

Following the imperialist model set by Washington, the ruling elites of many other capitalist nations have begun implementing similar counterintelligence strategies, imposing severe restrictions on socialist representatives and systematically dismantling networks that challenge the existing order.

Faced with escalating countermeasures, socialist intelligence officers must adopt more covert and sophisticated operational approaches to navigate heightened risks. This includes:

- Employing stricter security protocols in recruitment operations, ensuring that neither the recruiter nor the recruit is compromised.
- Utilizing deep-cover operatives and indirect recruitment methods, reducing exposure to hostile counterintelligence efforts.
- Maximizing the use of non-attributable communication methods, such as dead drops, one-time pads, and encrypted messages.
- Leveraging ideological sympathizers within adversary security structures, penetrating counterintelligence agencies to preempt and neutralize hostile operations.

Despite increased repression, the necessity of recruitment remains unchanged. The intelligence services of socialist states must continue to identify, cultivate, and recruit new agents, adapting their tradecraft to meet the challenges of a more complex and dangerous operational landscape.

The complex intelligence and operational landscape in capitalist states presents significant challenges, shaping both the strategies and methodologies employed in identifying, assessing, and recruiting individuals for covert cooperation with the intelligence services of socialist nations. Heightened counterintelligence efforts, increased surveillance, and political repression necessitate adaptability and precision in intelligence operations, particularly in the delicate process of recruitment.

The structure of recruitment operations is further influenced by the presence of both "legal" and "illegal" KGB stations within the target country. When both exist, they complement each other, creating opportunities for strategic maneuvering in recruitment operations. Legal residencies, operating under diplomatic or commercial cover, provide a front for overt contacts and intelligence collection, while illegal residencies, staffed by deep-cover operatives, offer a secure fallback for sensitive operations. The success of recruitment efforts depends not only on the existence of these complementary networks but on their effective management, strategic prioritization, and coordination. The direction of recruitment work, the composition and balance of personnel within the residency, and the political awareness, operational acumen, and recruitment expertise of intelligence officers are all decisive factors in determining the overall effectiveness of agent acquisition.

The course and outcome of recruitment efforts are shaped by numerous factors related both to the individual being recruited and to the external conditions under which the recruitment takes place. The age, gender, nationality, religion, social standing, financial status, and family background of the potential recruit all play a direct role in shaping the basis for recruitment and determining the most effective strategy for securing cooperation. These attributes influence the psychological approach, the nature of the initial contact, and the selection of leverage points that will ensure compliance and long-term reliability.

The temperament of the target is another crucial variable. A gradual approach, involving ideological cultivation, psychological persuasion, and incremental exposure to intelligence work, may be most effective for some individuals. Others, by contrast, may respond more favorably to a direct recruitment proposal, presented as a clear-cut opportunity for secret collaboration. Understanding which approach best suits the personality and circumstances of the recruit is essential in determining the ultimate success of the operation.

The complexity of recruitment operations in adversarial environments demands a high level of adaptability, psychological insight, and operational precision. Intelligence officers must assess not only who is recruitable but also how best to recruit them, tailoring their approach to each target while ensuring maximum security and operational effectiveness.

The success of recruitment operations conducted by the intelligence services of socialist states within capitalist countries depends on the efficiency, skill, political awareness, and persistence of intelligence officers in identifying, assessing, and securing cooperation from suitable candidates. The ability to rapidly and effectively search for potential recruits, study their backgrounds, and, most importantly, discern their personal interests, ambitions, and motivations is essential. These factors determine whether an individual can be successfully persuaded, influenced, or pressured into clandestine collaboration and whether they will consciously and reliably carry out the tasks assigned to them.

Intelligence officers operating in capitalist states must possess the capability and resolve to navigate and overcome the obstacles imposed by enemy intelligence, counterintelligence, and law enforcement agencies. The hostile security environment demands a high level of operational discipline, tradecraft proficiency, and strategic adaptability to avoid detection while penetrating key sectors. To achieve this, intelligence officers must fully exploit the recruitment pool available in capitalist societies, identifying individuals within government institutions, industry, academia, military structures, and political movements who are vulnerable to recruitment or sympathetic to the socialist cause.

The ultimate goal is to establish a cohesive and effective agent network within the target country—one that is capable of executing a wide range of intelligence, counterintelligence, and influence operations in alignment with the objectives of socialist intelligence services. A well-structured and securely managed network not only facilitates the collection of vital intelligence but also ensures long-term strategic penetration of enemy institutions, allowing socialist states to anticipate adversary actions, counter hostile activities, and shape political and economic developments in their favor.

PART II

IDENTIFYING AND SELECTING TARGETS FOR RECRUITMENT

Fundamental Principles of Target Selection

The primary criterion in identifying and selecting individuals for recruitment is their actual ability to fulfill intelligence assignments. The intelligence value of a potential recruit is determined by their position, access, and influence within a target institution, as well as their ability to provide direct or indirect access to classified or strategically significant information.

The capabilities of a recruitment target may vary in scope and nature. Some individuals may possess direct access to intelligence objectives, such as those working within government agencies, military institutions, or scientific research centers. Others may have indirect access, serving as intermediaries through personal or professional connections. Effective recruitment of individuals from both categories ensures that the intelligence station can acquire the most comprehensive intelligence on a given target while maintaining operational flexibility and security.

Candidates selected for recruitment as auxiliary agents must, by virtue of their official position or occupational role, possess the ability to facilitate intelligence tasks. Such individuals may not necessarily have direct access to classified materials but can support intelligence operations by providing logistical assistance, maintaining secure communication channels, or offering cover for operational activities.

In intelligence operations, an individual's potential usefulness in any given field of operational work is referred to as their intelligence capabilities. These capabilities are not static but can be developed and cultivated over time.

In some cases, intelligence services recruit individuals who, at the time of recruitment, do not yet possess immediate access to intelligence objectives. In such instances, efforts are made to either create opportunities for them or actively facilitate their integration into key sectors. Intelligence officers may ensure that a recruited individual gains access to strategic positions through career advancement, marriage, or cultivated social and professional connections.

In these scenarios, it is essential to identify individuals whose social standing, financial position, personal relationships, or environment provide them with the potential to infiltrate target organizations. By strategically placing such individuals within critical institutions, intelligence services can gradually expand their operational reach, ensuring long-term access to valuable intelligence sources.

Once a suitable candidate with intelligence potential has been identified, the intelligence officer must devote full attention to conducting a thorough assessment of the individual. The officer must determine whether there are realistic prospects for securing their agreement to cooperate as an agent. A comprehensive study of the individual is required, focusing on their ideological leanings, personal circumstances, motivations, and vulnerabilities.

For example, an intelligence officer may establish contact with a mid-level official in the Ministry of Foreign Affairs of a capitalist state who has direct access to encrypted diplomatic communications. The intelligence value of such an individual is beyond question. However, possessing access alone does not automatically make them a viable recruitment target. The officer must determine whether there are pre-existing conditions that could facilitate the individual's willingness—or necessity—to collaborate with a socialist intelligence service.

If, during the course of interaction, it becomes evident that the official is a committed supporter of the capitalist system, devoted to serving their government, then it would be unrealistic to expect them to cooperate on political or ideological grounds. In contrast, if the individual expresses disillusionment with capitalism, harbors doubts about its future, or even believes that socialism is the inevitable successor to the capitalist system, then they may be recruited on ideological and political grounds.

In some cases, recruitment opportunities arise not from ideology but from personal vulnerabilities and compromised situations. An official may be financially overextended, involved in illicit financial dealings, or entangled in unethical activities that, if exposed, could lead to disgrace, dismissal, or legal consequences. Under such conditions, the officer can offer protection, either by resolving the individual's financial problems or ensuring that damaging information remains undisclosed, in exchange for cooperation.

To make an informed decision on whether a foreign national can be recruited, the intelligence service must first obtain detailed and verifiable intelligence on the individual's political stance, financial situation, personal interests, psychological needs, and moral character. Only after a comprehensive analysis of these factors can intelligence officers determine which leverage points or inducements would be most effective in securing the individual's commitment to clandestine cooperation with the intelligence service of a socialist state.

A precise understanding of a target's political views, personal motivations, and desires—the factors that might compel them to cooperate with intelligence—is essential for correctly determining the basis of their recruitment. Without this, an intelligence officer risks selecting the wrong approach or miscalculating the target's willingness to engage in clandestine work.

In many cases, however, identifying the true motivations of a foreign target is difficult and complex. Many progressively minded individuals in capitalist societies take great care to conceal their political convictions, fearing that open expression of anti-capitalist or pro-socialist views may result in loss of employment, surveillance, persecution, or imprisonment. Others may hide illicit activities—acts of corruption, financial misconduct, personal indiscretions, or other offenses—because exposure would lead to dismissal, arrest, social disgrace, or the collapse of personal and professional relationships.

Given these realities, intelligence officers from socialist states must often conduct extensive intelligence-gathering operations to uncover the true political and personal orientations of potential recruits. This involves assessing their attitude toward ruling and opposition parties, their stance on the peace movement, their position on national liberation struggles, and their views on the imperialist policies and subversive activities of their own government. Additionally, an individual's financial situation, sources of income, personal interests, and moral character must be carefully analyzed to determine whether a viable recruitment basis exists and to establish the direction of further engagement that will lead to a successful recruitment effort.

Beyond ideological and material considerations, intelligence officers must also evaluate the personal qualities of the target, such as decisiveness, bravery, caution, or cowardice. These traits often determine how effectively a recruit can operate in a clandestine role and how reliably they will fulfill assigned intelligence tasks. However, such characteristics are rarely immediately visible and typically only emerge over time during the development process. Through systematic and sustained engagement, intelligence officers can refine their assessment, ensuring that the final recruitment approach is both precisely tailored and strategically effective.

Before proceeding with the development of a potential recruit, it is essential to determine whether they lack disqualifying personal traits that would make their involvement in intelligence work impractical or dangerous. Factors such as serious illness, mental instability, excessive talkativeness, or an inability to maintain secrecy are clear indicators that recruitment should not be pursued. While a full psychological and operational assessment is necessary before making a final decision, at the initial stage, it is sufficient to confirm that no obvious disqualifications exist.

Initial intelligence about a person—information that provides insight into their potential usefulness, their possible recruitment basis, and their business and personal characteristics—is referred to as a tip. In intelligence terminology, this term is often extended to refer to the individual themselves once such preliminary information has been gathered.

Techniques and Methods for Identifying Potential recruits (Lead Acquisition)

The process of identifying and selecting individuals for recruitment at penetration targets is carried out through a combination of agent-based intelligence collection and open-source research. Intelligence officers systematically study institutions, organizations, and their personnel through existing agent networks and legally available sources of information.

Legal sources of intelligence on targets and their personnel include:

- Official publications issued by government institutions and state agencies.
- Informational and reference materials that provide insight into the structure, staffing, and functions of the target institution.
- Press coverage, including media reports that highlight influential individuals within the target.
- Personal relationships and direct observations made by intelligence officers.
- The resources and connections available through diplomatic, trade, and cultural representations of socialist states in capitalist countries.

This open-source intelligence research plays a particularly critical role in newly established or developing residencies, where agent networks are still being built and direct access to key targets is not yet available. In such cases, intelligence officers must rely heavily on publicly available data to map out the organizational structure, personnel hierarchy, and potential vulnerabilities of the target institution.

To enhance their understanding of key institutions, intelligence officers leverage their personal acquaintances and professional contacts—not only those with direct connections to the target but also those with indirect access that can provide secondary insights. A well-placed contact at one government agency, research institute, or corporation may possess valuable knowledge about other institutions, allowing intelligence officers to cross-reference intelligence and uncover new recruitment opportunities.

By systematically combining agent-supplied intelligence, legal research, and social contacts, intelligence officers can effectively identify, assess, and prioritize potential recruitment targets, ensuring a steady pipeline of new assets for intelligence operations.

During the process of intelligence gathering, information naturally accumulates on a large number of individuals connected in some way to targets of interest. From this pool of names, intelligence officers carefully select those with potential intelligence value and begin developing recruitment strategies tailored to them. The process of identifying suitable candidates and mapping out possible recruitment approaches forms the foundation of agent network expansion.

The information that determines whether a person is a viable recruitment prospect, commonly referred to as a "tip," can originate from multiple sources. Agents who are already working within an intelligence network may provide recommendations based on their access and observations. Individuals who are unwittingly providing intelligence—those used "in the dark"—or those who maintain confidential yet informal relationships with intelligence officers may also reveal valuable insights about potential recruitment targets. Additionally, intelligence services may receive leads from their central intelligence apparatus, which processes and disseminates intelligence reports from multiple sources. Finally, intelligence officers themselves, through their personal contacts, professional connections, and legal sources of information, can identify and assess individuals who may be of interest for recruitment.

A practical example illustrates how intelligence officers generate tips through direct observation and strategic social engagement. An intelligence officer was tasked with recruiting agents capable of both reporting on and influencing the policies of the Social Democratic Party in the target country. To accomplish this, he began studying the party's leadership using official records and expanding his network of personal contacts within political circles. During this process, he identified a prominent party figure, known as "Laur," and arranged an opportunity to meet him at a public event, in this case, a film screening.

Through both direct interaction and reports from agents, the intelligence officer learned that Laur was sympathetic toward the Soviet Union and other socialist states. He was dissatisfied with the policies of the right wing of his own party but concealed his views out of concern for his professional position. This assessment confirmed Laur as a recruitment prospect, and he was subsequently placed under development. His recruitment was ultimately secured based on his political sympathies for socialist countries, with additional leverage provided by his anti-American sentiments and financial incentives, which reinforced his willingness to cooperate.

While various intelligence sources can contribute valuable information, the most critical source for acquiring documents and classified materials remains the agent network itself. Through effective recruitment and network expansion, intelligence services ensure a steady flow of sensitive information essential to their strategic objectives.

In the current intelligence-operational environment, particularly within Anglo-American bloc countries, restrictions on the movement and interactions of socialist state citizens have intensified. Diplomatic, trade, and cultural representatives from socialist nations face strict surveillance, limited access to local populations, and constant counterintelligence scrutiny. These conditions severely constrain intelligence officers' ability to independently identify and assess potential recruits, making the role of agents in generating leads increasingly vital.

When obtaining recruitment leads from an agent, an intelligence officer must exercise extreme discretion to avoid exposing their specific interests. Direct inquiries should be avoided, as they may arouse suspicion or lead to an agent drawing unwanted conclusions about intelligence priorities. Instead, it is advisable to issue a general directive instructing the agent to compile a broad list of their acquaintances and contacts, providing basic descriptions of their relationships, backgrounds, and professions. Over time, the officer should request updates and clarifications, ensuring that new connections are documented while outdated ones are removed. This method allows for a gradual and systematic identification of potential recruitment targets without alerting the agent to intelligence priorities.

This approach is especially critical when dealing with newly recruited or untested agents whose reliability and tradecraft skills have not yet been fully assessed. A compromised or reckless agent may inadvertently expose an intelligence operation by demonstrating suspicious curiosity, making direct inquiries, or approaching potential targets too aggressively.

Even within a loyal and well-trained agent network, careful verification of tips remains essential. Experienced agents who are committed to their tasks will often proactively suggest individuals they believe may be of recruitment interest. However, even the most reliable agents are not immune to error. Their assessments may be influenced by personal bias, misjudgment of an individual's true loyalties, or an overestimation of a contact's intelligence potential. For this reason, all tips—whether from seasoned operatives or from agents whose trustworthiness is still in question—must undergo rigorous verification and analysis before being acted upon.

In some instances, foreign intelligence services themselves unintentionally create opportunities for gathering leads by facilitating access to key social circles. This can occur through sponsored initiatives that grant agents opportunities to mingle with high-value targets, such as funding exclusive social clubs, fashion studios, political salons, or elite gatherings. Intelligence services can leverage these openings to insert their operatives, enabling broader access to potential recruits.

A concrete example illustrates this method in action. An intelligence officer was dispatched to a capitalist country with the objective of establishing an agent network within diplomatic circles. Upon arrival, he faced a significant operational challenge: his official cover role provided no access to foreign diplomats, and his only local asset was an aging former journalist who had lost most of his influential connections. In order to gain access to diplomatic figures and begin identifying recruitment prospects, the intelligence officer had to strategically engineer social and professional inroads into this closed environment.

This example underscores the reality that successful lead generation often requires creative adaptation to local conditions. Whether through leveraging existing agent connections, cultivating indirect access, or exploiting foreign-sponsored networking opportunities, intelligence officers must employ flexible and nuanced tradecraft to penetrate target circles and acquire recruitment leads without compromising security or arousing suspicion.

The intelligence officer determined that the agent had no immediate means of accessing diplomatic circles. However, he also discovered that the agent had previously published a small magazine, the editor of which was a well-connected Frenchman known in high-society circles. Recognizing this as a potential avenue for access, the intelligence officer instructed the agent to locate the Frenchman and determine whether he would be willing to serve as the editor of a newly established high-society diplomatic magazine. The agent successfully re-established contact and secured the Frenchman's agreement.

A publishing plan was drawn up, with two primary staff members: the agent as the publisher and the Frenchman as the editor. Within a month, a professionally designed magazine was launched, and within six weeks, it had secured subscriptions from embassies, diplomatic missions, consulates, and numerous aristocratic figures within the host country. The agent used this newfound legitimacy to print business cards identifying himself as the "Director-Publisher of the Diplomatic Magazine." This professional title granted him immediate access to diplomatic circles, allowing him to attend receptions, establish contacts, and interact freely with embassy personnel. Through this initiative, the agent gained significant opportunities for generating recruitment leads.

Each operational measure undertaken to obtain recruitment leads must be carefully tailored to the intelligence and operational climate of the specific country. The success of such initiatives depends on an intelligence officer's deep understanding of local conditions and their ability to employ creativity and adaptability in overcoming access barriers.

Citizens of socialist states stationed in capitalist countries, particularly high-ranking personnel within embassies, diplomatic missions, trade delegations, and other official representations, are also valuable sources of intelligence. Their professional responsibilities place them in frequent contact with local citizens, allowing them to gather critical information on potential recruitment targets. These individuals, through both direct interaction and indirect observations, can provide intelligence officers with valuable insights into the recruitment landscape.

Within the central intelligence offices of socialist states, where intelligence reports from all residencies operating in capitalist countries are compiled and analyzed, officers reviewing archival and current intelligence files occasionally uncover valuable leads. These findings often emerge from cross-referencing past intelligence materials with ongoing investigations, allowing the identification of individuals who may be viable recruitment targets.

Progressive figures within bourgeois states, when interacting with representatives of socialist institutions abroad, sometimes disclose information about their acquaintances or directly suggest individuals who may be sympathetic to socialist causes. Such exchanges occur during discussions of political and social issues, particularly in moments of ideological alignment. However, when receiving tips from such individuals, intelligence officers must exercise heightened vigilance, as counterintelligence agencies routinely monitor known progressive figures and may deliberately expose agents to them in an attempt to infiltrate or compromise socialist intelligence operations.

As in all intelligence activities, lead generation must be conducted with utmost discretion to avoid unintended consequences. One of the primary concerns is the potential harm that intelligence work may inflict upon progressive organizations. Special caution must be exercised when dealing with progressive individuals who hold leadership positions within democratic organizations. Before deciding to initiate recruitment, intelligence officers must carefully evaluate the balance between operational benefits and political risks. While recruiting such individuals may provide valuable intelligence, exposure could result in severe political fallout, damaging both the targeted individual and the broader movement they represent.

A fundamental principle that must always be upheld is that the intelligence services of socialist states must not recruit members of communist and workers' parties in capitalist countries. Such an act would endanger not only the individual communists involved but also the party as a whole, as exposure would provide imperialist authorities with justification to suppress and dismantle these organizations. Recent history has shown that Western intelligence agencies actively employ provocations and entrapment tactics to fabricate espionage accusations against communists, using such claims as a pretext for legal persecution, defamation campaigns, and broader efforts to undermine the legitimacy of socialist states and their diplomatic representatives.

Despite these challenges, the intelligence residencies of socialist states operating in capitalist countries possess extensive opportunities for obtaining recruitment leads. However, the effectiveness of this work depends on its careful execution, strict adherence to security protocols, and systematic long-term planning. The process of identifying, evaluating, and developing potential recruits must be covert, deliberate, and highly structured, ensuring that every opportunity is maximized while minimizing operational risks.

Assessment and Verification of Recruitment Leads

Tips obtained from agents and other intelligence sources rarely contain all the necessary details to determine with certainty whether a given individual possesses the required intelligence capabilities, whether a viable recruitment basis exists, and whether they have the necessary personal qualities to function effectively as an agent. Moreover, initial intelligence on a target is often incomplete, misleading, or outright false due to bias, misinformation, or deliberate deception. As a result, a tip alone is insufficient to justify active recruitment efforts without further verification.

To ensure that recruitment targets are properly vetted, intelligence officers must systematically analyze, cross-check, and corroborate leads. Given the presence of hostile agents, informants, and unreliable sources within a recruitment pool, rigorous scrutiny is essential to filter out compromised or strategically useless individuals before proceeding with any recruitment attempt.

The verification of leads is conducted through multiple intelligence-gathering methods. Additional information is collected through the agent network, as well as from citizens of socialist states who have been recruited in capitalist countries. These sources can provide supplementary biographical and professional insights, helping to refine assessments of a target's access, security risk, and suitability for recruitment.

In parallel, background checks and intelligence profiling are conducted at the target's place of residence and employment. This may involve external surveillance, discreet inquiries through third-party contacts, and the analysis of publicly available materials, such as newspapers, magazines, reference books, business directories, and other publications. Technological means, including operational surveillance tools and classified intelligence databases, are also employed to assess patterns of behavior, financial activities, and interpersonal connections that may indicate either vulnerabilities or risks.

Direct engagement is another essential method of assessment. When feasible, an intelligence officer, a group leader, or a trusted agent may be tasked with establishing personal contact with the target in a manner that does not raise suspicion. This contact serves as an opportunity to evaluate the individual's reactions, ideological inclinations, temperament, and possible openness to recruitment.

Finally, all intelligence on a potential recruit is cross-checked against the records of the central intelligence apparatus, ensuring that no contradictory or compromising information exists within existing files. This final step minimizes operational risks, preventing wasted efforts on undesirable, unreliable, or compromised individuals.

Through this methodical, multi-layered verification process, intelligence officers can accurately determine whether a target is suitable for recruitment, ensuring that only individuals with genuine intelligence value, a viable recruitment basis, and the necessary operational qualities are actively pursued.

Agents assigned to study and verify recruitment leads should, as a rule, remain unaware of the intelligence service's actual interest in the individual being investigated. To maintain operational security, the tasks given to agents for gathering information on a target must be disguised under unrelated pretexts. This approach ensures that the agent does not suspect that they are contributing to a recruitment effort, reducing the risk of compromising the operation due to speculation, indiscretion, or counterintelligence infiltration.

This method, however, complicates and prolongs the verification process, as intelligence officers must devise creative and indirect ways to collect the necessary data without revealing their true objective. Tasking agents in this manner requires ingenuity and precision, as well as an ability to maintain the plausibility of the cover story under which the inquiry is conducted.

In certain situations, exceptions to this rule may be made. The decision on whether to conceal or disclose the intelligence service's recruitment interest in a target depends on multiple factors, including the level of trust in the agent, the agent's role in the recruitment process, and their personal reliability. If an agent is directly involved in preparing the recruit or will be acting as the primary recruiter, it may be necessary to inform them of the true purpose of the assignment. However, even in these cases, intelligence officers must carefully control the flow of information, ensuring that only essential details are shared.

There are also instances where it is impossible to assign an agent to study a target without revealing some indicators of intelligence interest. When circumstances require an openly assigned task, intelligence may proceed with it. The primary advantage of this approach is that a fully informed agent can be more effective, applying independent judgment and creativity in their interactions with the target. They will understand what aspects to focus on, what specific intelligence to extract, and which methods will be most effective in obtaining the required information.

However, in cases where an individual is of exceptional intelligence interest, it is often inadvisable to disclose the ultimate objective—even to the most trusted agents. Such high-priority recruitment targets demand the highest level of secrecy, ensuring that the target remains unaware of being under assessment and that no leaks occur within the intelligence network. This level of discretion is critical in cases where even minor exposure could compromise an entire recruitment scheme.

In certain operational circumstances, external surveillance and background investigations may be required to further evaluate a recruitment lead. External surveillance is particularly useful in identifying the personal and professional connections of the target, as well as assessing their behavior outside of their workplace and home environment. A home installation, or domestic surveillance, provides critical insights into the target's financial situation, family dynamics, household relationships, and social circle. Additionally, such monitoring can yield important data on the target's political views, ideological leanings, and potential vulnerabilities.

By integrating covert intelligence collection, strategic agent tasking, and direct operational measures, intelligence officers ensure that all recruitment leads are thoroughly vetted before any approach is made. The careful balance between secrecy and information control remains fundamental to protecting both intelligence operations and the long-term viability of recruited agents.

Intelligence officers routinely examine recruitment leads through printed materials, including newspapers, magazines, reference books, and other publicly available publications. This process involves systematically collecting and analyzing articles that provide insights into the background, professional activities, public reputation, and personal characteristics of the individual under study. Such sources may offer valuable contextual information on the target's political stance, ideological leanings, affiliations, and influence within relevant circles.

In addition to open-source research, all recruitment leads must be cross-checked against the intelligence records of the central intelligence apparatus. Regardless of how comprehensive the information obtained through agents, external surveillance, home installations, and printed publications may appear, intelligence officers are required to conduct a final verification using the classified archives and operational records of state security agencies. These records consolidate intelligence from multiple sources over extended periods, providing a detailed and historically comprehensive profile of the target.

Such a background check may reveal critical intelligence that would otherwise go unnoticed. The investigation may uncover that the target has connections to hostile intelligence services, has engaged in suspicious activities in the past, or possesses unfavorable biographical details that raise security concerns. Conversely, the verification process may establish that the individual has already been recruited or studied by another branch of our intelligence services, requiring coordination before proceeding further.

Direct interaction between intelligence officers and the subject of interest plays an essential role in the recruitment assessment process. Personal engagement—whether through casual interactions, professional exchanges, or staged social encounters—enables an intelligence officer to observe the target's behavior, test their reactions, and validate or challenge existing intelligence. Through such direct contact, intelligence officers can confirm whether an individual is genuinely a viable recruitment target or if they are unreliable, deceptive, or potentially compromised.

Personal engagement also serves an important counterintelligence function. By directly interacting with the target, an intelligence officer can identify individuals who may be spreading disinformation, seeking personal gain, or acting as provocateurs on behalf of adversary intelligence services. At the same time, previously unknown qualities and characteristics of the target—which agents may have overlooked or failed to recognize as significant—can be independently assessed and documented.

For an intelligence officer, establishing personal contact with a recruitment lead must be carefully orchestrated to ensure that it appears entirely natural and circumstantial. The subject should never suspect that they are being deliberately approached as part of an intelligence assessment. The method of introduction and interaction is determined by multiple factors, including the intelligence officer's official position, whether they are operating under legal or illegal cover, and the target's professional and social standing. The success of this process relies on a precise understanding of the target's lifestyle, behavioral patterns, habits, interests, and psychological traits, allowing the intelligence officer to blend seamlessly into the environment while conducting a covert evaluation.

Given these considerations, the task of personally acquainting oneself with a subject under study should be assigned to the intelligence officer who can do so most naturally and effortlessly within the given circumstances. The initial contact must appear organic and should not raise suspicion. A variety of pretexts may be employed, including official business engagements, social receptions, or shared interests such as sports, hunting, fishing, or other leisure activities. These settings provide plausible opportunities for interaction while allowing the intelligence officer to assess the individual in a relaxed and seemingly unscripted environment.

In addition to direct personal engagement, letters of introduction and recommendation serve as valuable tools for facilitating contact and establishing relationships with individuals of recruitment interest. These letters may be provided by relatives, friends, or acquaintances of the subject, whether they reside in socialist or capitalist countries. When used effectively, such letters can lend credibility to an introduction and significantly ease the process of building rapport.

The content and format of these letters of recommendation can vary widely depending on the circumstances and the nature of the relationship between the author and the recipient. In some cases, the letter may be a brief note or business card, simply indicating that the bearer seeks assistance in establishing correspondence. Other letters may contain specific requests, such as helping the bearer find a broker, purchase property, or become acquainted with influential individuals. More elaborate letters might include personal appeals, referencing past acts of financial assistance, life-saving interventions, or other favors rendered by the bearer, with the author requesting that the recipient extend support in return.

Whenever possible, these letters should be handwritten by the actual author to enhance authenticity. However, when circumstances prevent this, the letter may be fabricated by intelligence services. In such cases, extreme care must be taken to ensure accuracy in every detail. The intelligence officer responsible for drafting the document must first determine whether the recipient is familiar with the author's handwriting, writing style, and personal background, incorporating these details into the forged letter to maintain plausibility. Failure to do so could result in immediate suspicion, jeopardizing not only the intelligence operation but also the security of the personnel involved.

By carefully selecting the most credible and inconspicuous means of introduction, intelligence officers can effectively initiate contact with high-value targets while minimizing operational risks. Whether through direct social engagement or indirect introductions facilitated via letters of recommendation, the key to success lies in seamless integration into the target's social or professional circles, ensuring that the approach appears entirely natural and uncontrived.

The ability to thoroughly study and assess individuals of intelligence interest depends significantly on the official status of intelligence officers operating in socialist countries. Intelligence officers embedded in illegal residencies, who assume the identities of local residents or citizens of third countries, generally operate under more favorable conditions. Their status allows them greater freedom of movement, enabling them to travel between cities, establish acquaintances, and engage with a diverse range of individuals without drawing suspicion. These advantages greatly enhance their ability to identify, observe, and assess potential recruits, making them particularly effective in the early stages of intelligence work.

Intelligence officers stationed in legal residencies, such

as embassies, trade missions, and cultural institutions, also have opportunities to study leads within the limits of their official roles. However, due to the routine surveillance imposed by counterintelligence agencies, their movements and interactions are closely monitored. Officers in legal cover positions must therefore develop the skill and resourcefulness necessary to circumvent these restrictions, ensuring they can discreetly establish and maintain contact with recruitment leads without arousing suspicion or compromising their operational security.

As recruitment leads are studied, the majority are ultimately discarded. This process of elimination is both natural and necessary, as preliminary assessments frequently reveal that many individuals initially considered for recruitment lack the intelligence capabilities required for useful cooperation and show no realistic potential for gaining such access in the future. In other cases, further investigation determines that the prospective recruit lacks a viable recruitment basis, meaning they have no strong ideological, financial, or personal motivation that could serve as leverage for recruitment. Additionally, certain leads may exhibit undesirable personality traits or security vulnerabilities that make them unsuitable for intelligence work.

Those leads who, through preliminary study and verification, are confirmed to possess intelligence value—either through existing access or a clear path to acquiring it—and who demonstrate a viable recruitment basis along with the necessary personal characteristics for clandestine activity are taken into active development. At this stage, formal recruitment files—known as recruiting dossiers—are opened on them within the central intelligence apparatus. These files serve as the foundation for structured recruitment planning, ensuring that each prospect is methodically assessed, cultivated, and ultimately converted into a reliable intelligence asset for the socialist intelligence services.

For a person to be considered a viable recruitment target, they must meet several key criteria. First, they must already possess intelligence capabilities—meaning direct or indirect access to valuable information—or have the realistic potential to acquire such access in the future. Without this fundamental qualification, a recruit would hold no operational value to intelligence services.

Second, the individual must have beliefs, aspirations, interests, needs, or desires that can serve as the basis for recruitment. This basis may already exist, such as ideological alignment with socialist principles, financial hardship, personal grievances against the target government, or a desire for career advancement. Alternatively, it may be artificially created over time through psychological influence, controlled exposure to ideological materials, or the strategic exploitation of vulnerabilities.

Third, the recruit must demonstrate personality traits that indicate they are capable of fulfilling intelligence tasks—either immediately or after a period of conditioning and influence by intelligence officers. This includes qualities such as discretion, adaptability, reliability, and a capacity for secrecy. If an individual lacks these characteristics at the outset, an intelligence officer must assess whether they can be gradually shaped into an effective operative through training and psychological reinforcement.

Once a recruitment prospect has been fully assessed and deemed viable, all materials related to their intelligence value, motivations, and personal characteristics are sent to the central intelligence office. There, a final review and approval process takes place. If no security concerns or objections arise, the recruitment project is formally authorized, and a case file is opened on the target.

At this stage, the residency retains only the essential details necessary for managing further recruitment efforts, while the central intelligence office takes over formal documentation and strategic oversight of the operation. This ensures that sensitive intelligence records remain compartmentalized and that the recruitment process is conducted with maximum security and efficiency.

Maintaining Operational Security and Counterintelligence Awareness During Recruitment Target Evaluation

In the process of identifying, studying, and verifying recruitment leads, intelligence officers must exercise the highest level of vigilance and adhere to strict operational secrecy. Any lapse in security protocols when searching for and evaluating leads creates the risk of enemy counterintelligence infiltration, potentially exposing operations, compromising agent networks, and jeopardizing intelligence activities.

Experience has shown that leads presented to us as having strong "intelligence capabilities" and a clear "recruitment basis" often appear moderately promising rather than overtly ideal. This is a deliberate tactic employed by hostile counterintelligence agencies, which understand that an overly valuable or overly easy recruitment prospect might arouse suspicion. Conversely, if a lead is entirely unremarkable or lacks any strategic access, it may not attract interest at all.

For this reason, hard setups orchestrated by enemy counterintelligence are designed to be difficult to detect through surface-level observation alone. Intelligence officers must exercise extreme vigilance and conduct a series of intelligence and operational verification measures to identify potential traps before proceeding with recruitment efforts.

Every aspect of how a lead was obtained must be critically analyzed. Intelligence officers must thoroughly examine the circumstances surrounding the tip, scrutinizing whether there are hidden counterintelligence setups embedded within the recruitment lead. Any anomalies or inconsistencies in the source of the tip, the background of the target, or the manner in which the information was introduced should raise immediate concerns and trigger further verification efforts.

To prevent compromise or enemy counterintelligence countermeasures, intelligence officers must maintain absolute secrecy when receiving leads and conducting verification activities. Any carelessness, failure to operate covertly, or deviation from security protocols could alert adversary services to the intelligence service's specific recruitment interests. If counterintelligence agencies register and track this interest, they may actively work to disrupt recruitment efforts, introduce controlled setups into the agent network, or organize a provocation during the recruitment conversation.

The consequences of a security breach at this stage can be severe, leading to the loss of valuable assets, exposure of intelligence officers, and potential long-term damage to operational effectiveness. To mitigate these risks, every action taken in lead identification, study, and verification must be conducted with precision, discretion, and strict adherence to intelligence tradecraft principles.

Maintaining absolute secrecy in the verification of recruitment leads is critical from the very outset. Any lapse in security during the initial study and verification process can compromise an otherwise viable lead, forcing intelligence officers to abandon recruitment efforts due to premature exposure. If a lead is disclosed too early, continuing to work with them poses a constant security risk, as there would always be uncertainty about whether they have been turned into a counterintelligence provocateur.

When studying targets and their personnel, whether through agent reports or legal intelligence sources, it is imperative that enemy counterintelligence never detects intelligence interest in a specific individual. The moment an adversary service becomes aware of socialist intelligence efforts targeting a particular person, it will take immediate countermeasures to block recruitment attempts. This may include placing the individual under surveillance, restricting their access to sensitive information, or attempting to manipulate them into becoming a planted informant who will mislead or expose intelligence operations.

The process of lead verification must be both systematic and efficient. Intelligence officers must skillfully combine and apply all available resources—both from illegal and legal residencies—to conduct thorough yet timely evaluations. Prolonged assessments can increase the risk of exposure, allowing enemy intelligence services the opportunity to identify and neutralize recruitment efforts. By correctly leveraging various intelligence-gathering methods, intelligence officers can accelerate the verification process, ensuring that viable leads are developed into active assets before adversary countermeasures take effect.

The correct application of intelligence tradecraft, flexibility in approach, and creativity in operational execution enables intelligence officers to effectively counter the efforts of enemy intelligence and counterintelligence agencies. A well-structured approach ensures that provocateurs, disinformers, and double agents are identified in time, preventing infiltration while allowing genuine recruitment prospects to be developed securely.

A practical example demonstrates the importance of checking central intelligence records to identify enemy provocateurs before a recruitment effort is compromised.

The Soviet embassy in the capital of a European state was occasionally visited by a man known as "Gopper," the editor and publisher of a bourgeois magazine. During conversations with embassy staff, Gopper frequently expressed strong pro-Soviet sentiments, spoke approvingly of the USSR and other socialist nations, and was openly critical of American foreign policy. Whenever the opportunity arose, he hinted at his willingness to assist the Soviet Union, claiming that he wanted to contribute to strengthening the cause of peace, democracy, and socialism.

Gopper's access to diplomatic circles, his broad network of contacts within political and social elites, and his measured approach in presenting himself piqued the interest of the Soviet residency. He was not overly persistent, which made him appear more credible. Given his connections and potential intelligence value, the residency decided to assign an intelligence officer to establish and manage contact with him. The officer was chosen because, given his official position, he could maintain a legitimate professional relationship with Gopper without raising suspicion.

At a reception at the Hungarian diplomatic mission, the intelligence officer arranged a follow-up meeting with Gopper at the Soviet embassy. During this meeting, Gopper provided personal details about himself, reiterated his support for Soviet foreign policy, and expressed outrage at a recent speech by a reactionary member of parliament who had allegedly founded a new right-wing organization. Gopper volunteered to gather more information on this group, and the intelligence officer, while remaining cautious, approved the initiative as a test.

Not long afterward, Gopper arrived at the embassy unannounced. He handed over a dossier containing details about the leadership of the reactionary organization, along with confidential internal materials published by the group. He then took his offer further, stating that he was willing to infiltrate the organization himself if it would be beneficial to Soviet intelligence. Emphasizing his unwavering support, he assured the intelligence officer that he could fully count on him, as he believed that a new war was being prepared and only the socialist countries could prevent it.

The intelligence officer expressed gratitude for Gopper's willingness to assist and agreed to another meeting. However, in accordance with standard security protocols, the residency immediately reported the development to the central intelligence apparatus and requested a background check on Gopper using operational records. Throughout his interactions, the intelligence officer remained passive, never displaying enthusiasm or urgency, and carefully conducted his conversations under the cover of an embassy official merely studying political affairs.

The central intelligence apparatus completed the background check, and the results exposed Gopper as an agent of German intelligence, with a history of spying against the USSR dating back to 1933. His entire presentation as a Soviet sympathizer had been a deception, likely an attempt to penetrate Soviet intelligence networks on behalf of a hostile counterintelligence agency.

Upon learning of this, the central intelligence apparatus immediately instructed the residency to sever all contact with Gopper under a plausible pretext, ensuring that the decision would not arouse suspicion. The order was executed without delay, effectively preventing an enemy counterintelligence infiltration attempt.

This case demonstrated the critical importance of rigorous lead verification before advancing any recruitment efforts. Without the systematic background check, the intelligence officer might have unknowingly compromised the integrity of the agent network. The enemy's attempt to introduce a controlled informant was neutralized, reinforcing the necessity of vigilance, thorough investigation, and strict adherence to security protocols in all intelligence operations.

When an agent proactively suggests a recruitment lead, the intelligence officer must always carefully examine why and under what circumstances the agent is making this recommendation. It is essential to remain aware that enemy intelligence and counterintelligence agencies actively seek to expose their own controlled agents to socialist intelligence services by exploiting individuals suspected of having ties to socialist representatives. In many cases, the individual being used as a setup is unaware that they are serving as a provocation tool for the adversary's counterintelligence service.

Similarly, an agent who provides a lead may be acting with complete sincerity, unaware that enemy counterintelligence, having identified their connection to socialist intelligence, is using them to introduce a planted informant into the agent network. Such scenarios are well-documented in operational practice, as illustrated in the following case.

A socialist intelligence residency had been tasked with studying personnel within the Ministry of Foreign Affairs of a capitalist country to identify a recruitment candidate who could provide insights into the government's internal policymaking. The resident instructed officers to propose measures for obtaining leads on individuals of intelligence value. Among those assigned this task was an intelligence officer handling the agent "Mars," who worked within the Ministry of Foreign Affairs and was considered a trusted and reliable source. The officer instructed Mars to monitor his professional contacts and report on individuals who might be of interest for recruitment.

After some time, Mars provided the intelligence officer with a document that had passed through his hands. This document revealed that the Deputy Minister of Foreign Affairs, identified here as "Sachs," had been forcibly removed from his position at the insistence of the British. According to the document, Sachs had been accused of aligning with nationalist elements and pursuing an anti-British foreign policy, leading British authorities to engineer his dismissal. Additionally, the document suggested that Sachs had a long history of collaboration with British intelligence and had advanced his career with their support.

The intelligence residency found this information operationally valuable. If accurate, it meant that Sachs had deep institutional knowledge, extensive connections within the Ministry of Foreign Affairs, and potential access to classified government decisions. Furthermore, his apparent break with British intelligence could be leveraged as a political grievance, providing a basis for recruitment. The officers, however, failed to critically assess the broader implications of the intelligence. They did not consider whether British intelligence would allow a former high-level asset to operate freely against them, particularly in a country where British influence was strong and intelligence operations were deeply embedded.

During a reception event, the intelligence officer arranged an initial meeting with Sachs, where Sachs confirmed that he had been removed from his post due to British intervention. Based on this, the intelligence officer began developing Sachs as a recruitment target. Eventually, Sachs was formally recruited into the socialist intelligence network.

At a later meeting, Sachs provided a list of individuals connected to British intelligence. One of the names on this list was "Leon," a trusted agent already working for the socialist intelligence service. This immediately raised suspicions. Further investigation revealed that Sachs' removal from office had been a staged operation orchestrated by British intelligence. The goal had been to build an artificial reputation for Sachs as a supposed adversary of the British, thereby making him appear attractive to socialist intelligence services.

British intelligence had executed this deception by exploiting the agent Mars, whom they suspected of working with socialist intelligence. Without Mars realizing it, they planted the fabricated document, ensuring that socialist intelligence would develop interest in Sachs and ultimately recruit him—thus inserting a controlled British asset into the agent network.

If the intelligence residency had critically examined why the British had treated Sachs so harshly, they might have been more cautious. A deeper analysis of his actual ties to British intelligence could have exposed the deception before recruitment occurred.

To prevent enemy infiltration of socialist intelligence networks, officers must remain highly skeptical of recruitment leads, particularly those provided by agents whose reliability could be compromised. Intelligence officers must be attentive to the circumstances under which a lead is received, support their agents in understanding the broader operational environment, and closely monitor their actions to ensure they are not being manipulated by enemy counterintelligence. Only by maintaining strict verification procedures, continuous assessment of agent reliability, and critical analysis of recruitment leads can intelligence officers successfully distinguish between genuine opportunities and enemy deception operations.

PART III

THE RECRUITMENT PROCESS AND ITS STRATEGIC COMPONENTS

Fundamentals of Recruitment Operations

Recruitment is fundamentally based on the material and spiritual needs of an individual, as well as their personal characteristics, which intelligence officers exploit to secure their cooperation in a clandestine manner.

Socialist intelligence work abroad distinguishes three primary recruitment bases: ideological-political, material, and moral-psychological. The ideological-political basis relates to a recruit's spiritual or ideological alignment with the socialist cause, while the material basis is driven by financial or economic incentives. The moral-psychological basis is broader, incorporating both spiritual and material elements. For instance, the threat of exposure—a classic recruitment lever—can affect both financial stability and personal integrity. Similarly, emotions such as jealousy, love, and hatred also fall under the moral-psychological category, as they shape a recruit's psychological motivations.

In most cases, recruitment is not based on a single factor but rather a combination of these three elements. It is common for ideological commitment to be reinforced by material incentives, or for material dependency to be linked to emotional and psychological vulnerabilities. Successful recruitment strategies often rely on identifying which combination of motivations will be most effective for a given target.

Recruitment Based on Ideological and Political Convictions

An ideological-political recruitment basis exists when a target sympathizes with the socialist bloc and possesses political views and interests that align—either partially or fully—with the goals of socialism and democracy. Their commitment to socialist ideals must be strong enough to serve as an incentive for clandestine cooperation with the intelligence services of a socialist country.

However, while shared political views and ideological alignment increase the recruitment pool, they do not automatically guarantee recruitment success. Many individuals who support socialist policies may hesitate or outright refuse to collaborate with intelligence services due to a misguided sense of patriotism, fear of exposure, or misconceptions about the role of socialist intelligence. Some may falsely believe that assisting a socialist intelligence agency would betray their homeland, while others may fear the personal and professional consequences of being exposed as an intelligence source.

It is not necessary for a recruit's entire worldview to align with socialist ideology. Often, it is sufficient to focus on a few key political issues where interests align. For example, a potential recruit may believe in peaceful coexistence between socialist and capitalist systems, support anti-war efforts, or oppose imperialist economic exploitation. These commonalities serve as an entry point for recruitment efforts, even if the individual does not fully embrace socialist ideology.

To successfully exploit these motivations, intelligence officers must first influence and strengthen the target's political convictions. The goal is to reinforce their sympathy toward socialist states, convincing them that assisting a socialist intelligence service is an extension of their existing political beliefs and contributes to a greater cause.

For instance, if a potential recruit is dissatisfied with American economic policies toward their country but is not actively engaged in resistance, an intelligence officer must persuasively highlight the exploitative nature of American policy. The recruit should be shown concrete examples of how the United States seeks to dominate their country's economy, reducing it to a source of imperialist profit. At the same time, they should be patiently and skillfully introduced to the peace-driven policies of socialist states, emphasizing the respect socialist nations have for both large and small countries and their willingness to fight against American expansionism.

When a recruit reaches the conclusion that neutrality is no longer an option and that they must actively resist imperialist aggression, their ideological-political recruitment basis is considered secured. However, intelligence officers must always adapt their approach to the specific circumstances, social standing, intellectual level, and personality of the recruit.

Socialist intelligence services are not only capable of strengthening existing political convictions but also of creating ideological commitment where none previously existed. This can be achieved by leveraging personal interests, cultural affinities, and external circumstances to shift the recruit's perception of socialist countries.

A recruit's passion for socialist literature, classical or modern music, technological advancements, or historical achievements can be subtly used to cultivate ideological alignment. Even an individual with no initial political leanings can be guided toward a favorable perception of socialist ideals, making them susceptible to recruitment on ideological-political grounds.

An illustrative example of how an ideological and political basis for recruitment can be created is the case of Schmidt, a prominent mathematician and engineer who, at the time of initial interest, did not hold the political convictions necessary for recruitment. Schmidt, the chief designer at an artillery plant, was an emigrant who had lost contact with his sisters living in the Soviet Union. Unlike many of his contemporaries, he had not been involved in émigré organizations and had instead focused on his engineering career. Through skill and technical expertise, he quickly advanced to a position of influence within the plant. However, he remained politically indifferent and harbored a distinctly negative attitude toward communism.

Recognizing his professional significance and the potential for ideological influence, intelligence officers decided to establish contact and identify elements in his worldview that could be leveraged for recruitment. The agent Suffrant, who had an established relationship with Schmidt, was tasked with introducing him to an operative from the Soviet trade mission under a neutral pretext.

Upon meeting Schmidt, the intelligence officer quickly assessed that his political understanding was shaped entirely by bourgeois newspapers, which had instilled in him a fear of communism. However, at the same time, Schmidt expressed deep hostility toward Germans, a sentiment that had developed during the First World War, when he came to associate them with the militarism that had once threatened his homeland.

Carefully and patiently, the intelligence officer engaged Schmidt in discussions about global politics, contrasting the foreign policy of socialist states with that of imperialist nations. Without pressuring him, the officer subtly introduced perspectives that countered his previous misconceptions, allowing Schmidt to gradually reassess his stance. Over time, the officer fostered a dynamic in which Schmidt increasingly sought him out for political discussions. The turning point in their relationship came when Schmidt, of his own accord, asked the officer for assistance in locating his sisters in the Soviet Union. The intelligence service arranged for contact to be reestablished, reinforcing the officer's credibility and deepening Schmidt's emotional connection to the Soviet state.

As their relationship progressed, Schmidt began to question his prior views. Exposure to new information, combined with the officer's skillful reinforcement of his preexisting hostility toward German militarism, led him to conclude that only the socialist nations were actively resisting its resurgence. The intelligence officer built upon Schmidt's patriotic feelings, guiding him toward the realization that collaboration with socialist intelligence was not just an option, but a necessity in the fight against imperialism. This ideological shift ultimately led Schmidt to voluntarily hand over the design for a new artillery projectile he had developed. Once this initial step was taken, he became increasingly involved in intelligence work, eventually transitioning into more serious activities in support of Soviet intelligence.

This case demonstrates that ideological and political alignment is not always a prerequisite for recruitment. Through careful influence, intelligence officers can shape the convictions of individuals who may have previously held apolitical views or even harbored opposition to socialism. By identifying and nurturing key emotional and ideological triggers, an intelligence service can transform an uncertain or resistant individual into a committed and reliable asset.

Regardless of how clearly the ideological and political basis for recruitment emerges during the study of a candidate or their development, intelligence officers must always refine and specify it when drafting a recruitment plan. It is essential to determine with absolute precision which ideological convictions of the candidate should be relied upon to secure their consent for clandestine cooperation. Without this careful preparation, even a seemingly promising ideological recruit may hesitate or fail to fully commit when approached.

Intelligence officers operating within legal residencies of socialist countries have optimal conditions for acquiring agents on an ideological-political basis. As official representatives of their state institutions, they have legitimate opportunities to identify and engage with individuals who are ideologically sympathetic to socialist policies. This process can take place under the pretext of professional obligations at official receptions, diplomatic gatherings, and formal meetings, where they can observe, assess, and gradually draw potential recruits into cooperation. Legal residencies also have significant flexibility in using their existing agent networks to study and cultivate targets for recruitment based on ideological and political alignment.

For intelligence officers working within illegal residencies, recruiting on an ideological and political basis is more complex and requires far greater discretion. Unlike their counterparts in legal cover positions, illegal intelligence officers must eventually reveal their intelligence affiliation to the recruit, making the process inherently riskier. Therefore, ideological and political recruitments carried out by illegal officers must be approached with exceptional caution, and only when there is high confidence in the recruit's commitment and discretion. To further mitigate risks, it is advisable that such recruitments take place in third countries, where local security services may have less interest in interfering, and where a failed recruitment attempt is less likely to result in direct exposure. A rejected recruit in a foreign country is also less likely to approach authorities with information about an illegal intelligence officer than they would in their own homeland.

Illegal residencies are particularly well-suited for ideological and political recruitment when carried out through agent recruiters, rather than direct engagement by intelligence officers. Using agents or agent networks acting under the cover of progressive organizations, intelligence officers can exploit the recruit's hostility toward the United States, the United Kingdom, or other imperialist powers. In such cases, every precaution must be observed. The connection between the ideological recruit and the illegal intelligence officer should be structured so that it is indirect, preferably facilitated through liaison agents or by arranging meetings in third countries to minimize operational risks.

The final phase of the recruitment process, the recruitment interview, may be conducted by intelligence officers dispatched directly from the central intelligence apparatus, or by officers and agents arriving from other countries for the specific purpose of finalizing the recruitment. This structured approach ensures that both legal and illegal residencies are fully capable of building a loyal and ideologically committed agent network within capitalist countries, strengthening the intelligence presence of socialist states.

Agents recruited on an ideological and political basis tend to be the most stable, proactive, and reliable assets. Their commitment is driven not by material incentives but by their deeply held convictions, making them less susceptible to betrayal, coercion, or wavering loyalty. Many ideological recruits have demonstrated exceptional dedication, willingly taking great risks, enduring hardships, and, in some cases, sacrificing their personal safety for the cause. This is particularly true for those who have experienced persecution under fascist regimes, social injustice, or national oppression at the hands of imperialist governments. Their personal histories and ideological convictions make them invaluable operatives, as they see their intelligence work not as a duty, but as a moral necessity in the struggle against imperialism.

Recruitment Based on Financial or Material Incentives

Recruitment based on financial or material incentives can be defined as the financial or economic needs of a target, where the urgency to satisfy these needs is strong enough to persuade them to cooperate with a socialist intelligence service, despite the potential risks and consequences imposed by their government. These material needs can arise from personal hardship, such as financial difficulties caused by illness, the burden of supporting a large family, the cost of children's education, or the necessity of securing financial stability for old age. Even individuals who appear well-off may develop material vulnerabilities due to a habit of extravagant living beyond their means, making them susceptible to financial incentives. Others are driven by pure greed, seeking ever greater wealth. This is particularly common among businessmen and industrialists, for whom financial gain is paramount. Some of these individuals actively seek profitable transactions with socialist trade institutions, presenting opportunities for intelligence services to exploit their ambitions for further enrichment.

Government officials in capitalist states, particularly those who have access to classified information, may also be vulnerable to recruitment on a material basis. Some are already willing to sell confidential state information to corporations for personal gain, making them more inclined to consider intelligence work as an extension of this practice. The capitalist value system, particularly in bourgeois societies, fosters a culture in which personal success and financial achievement are considered life's highest goals, often taking precedence over collective responsibility or loyalty to the state.

American culture, in particular, places a high value on financial prosperity as the primary measure of success. American sociologists such as Robert Merton and C. Wright Mills have observed that money is the ultimate indicator of status in the United States, and economic success is regarded as a fundamental aspect of an individual's identity. This cultural emphasis on wealth explains why a significant number of American agents have been recruited on a material basis, as financial incentives align with their deeply ingrained societal values.

Despite the effectiveness of recruiting agents through material inducements, such agents cannot be considered as reliable or committed as those recruited on ideological and political grounds. Agents motivated purely by financial rewards often view their intelligence work as a transactional arrangement, rather than as a commitment to a greater cause. For this reason, it is necessary to influence and educate these agents continuously, reinforcing their material dependence with ideological conditioning. By cultivating their awareness of the broader political struggle against imperialism, intelligence officers can strengthen the ideological foundation of their cooperation, making them less prone to defection or disloyalty.

Experience has shown that many individuals initially recruited solely for financial reasons can, through persistent political and psychological reinforcement, develop genuine loyalty to socialist intelligence services. Intelligence officers play a crucial role in this transformation by gradually shifting an agent's perspective from purely self-serving motives toward an awareness of the greater ideological and geopolitical stakes involved in their cooperation.

In addition to leveraging existing material vulnerabilities, intelligence services can actively create or exacerbate financial difficulties for potential recruits, ensuring their dependence on material assistance. Various operational measures can be employed to increase financial pressure on a target, making them more susceptible to recruitment efforts. Individuals with extravagant lifestyles and reckless spending habits can be manipulated into accumulating excessive debts, placing them in a position where they feel compelled to seek financial relief. In such cases, intelligence officers can offer material assistance in exchange for cooperation, creating an environment in which recruitment appears to be the only viable solution to the recruit's financial distress.

By strategically engineering financial dependency, intelligence services can turn economic pressure into a powerful tool for securing agent commitment. While material-based recruits may initially cooperate for self-serving reasons, careful management and continued psychological engagement can convert them into long-term intelligence assets, reinforcing their loyalty and minimizing the risk of compromise.

An example of how a material basis for recruitment can be created is the case of Fin, a cipher clerk at the American embassy in a capitalist country. The intelligence service of a socialist state became aware of him through an initial tip, which indicated that he was romantically pursuing Nona, the daughter of a well-known doctor in the host country. According to the intelligence obtained, Nona did not reciprocate Fin's affections, providing an exploitable weakness.

Further investigation into Fin's background revealed that he was loyal to the American government, ambitious, and persistent in achieving his personal goals. He also had a strong preference for dressing well and maintaining an appearance of success, which suggested a potential vulnerability to financial strain. Intelligence on Nona indicated that she was sociable, active in the peace movement, and enjoyed frequenting high-end restaurants and entertainment venues. Additionally, the intelligence service discovered that an experienced agent from the Grom residency had an existing relationship with Nona's father, providing a direct means of access to her.

The residency decided to first recruit Nona and then use her to manipulate Fin into financial distress, thereby creating a material dependency that could serve as the basis for his recruitment. The agent from Grom leveraged his acquaintance with Nona's father to gain her trust, and through a carefully cultivated relationship, successfully recruited her for intelligence work.

Once Nona was sufficiently studied and it was confirmed that Fin was genuinely in love with her, she was introduced to a residency operative who took over her handling and provided direct supervision of her intelligence activities. She was assigned the specific task of strengthening her relationship with Fin, gradually influencing his decisions and behavior. Through frequenting expensive restaurants, theaters, and organizing costly outings, she led Fin into accumulating significant financial burdens. As his expenses grew, she also subtly introduced him to a resident intelligence officer posing as a creditor, further entangling him in financial obligations.

As planned, Fin soon found himself in serious debt and, facing a financial crisis, sought an alternative source of income. Seeing the intelligence service of the socialist country as a way to solve his financial difficulties, he approached them and agreed to cooperate in exchange for financial support. The material dependency that had been carefully engineered was now successfully exploited, and Fin was recruited as an intelligence asset.

Cases of recruitment on a material basis are common, and in many instances, individuals themselves—motivated by financial difficulties—offer their services to intelligence agencies in hopes of compensation. However, material recruits must always be treated with caution, as they are more susceptible to counterintelligence manipulation. Some of these volunteers may be enemy setups, attempting to infiltrate the intelligence service on behalf of hostile agencies. Others, while genuinely seeking financial support, may prove to be unreliable under pressure. Because of these risks, agents recruited on a material basis undergo more rigorous and systematic verification than those recruited for ideological or political reasons. While some material recruits eventually develop loyalty through prolonged engagement and ideological reinforcement, they can never be considered as inherently stable or trustworthy as those who are ideologically committed.

Recruitment Based on Psychological and Moral Leverage

Recruitment based on moral and psychological leverage relies on an understanding of the candidate's moral character, psychological traits, and personal motivations. This approach leverages factors such as ambition, envy, jealousy, vindictiveness, attraction, affection, personal disposition, sympathy, love, or hatred to induce cooperation. Additionally, intelligence officers may exploit compromising facts about a candidate's behavior, particularly instances where they have violated the moral norms or legal standards of the society in which they live.

When applying this basis for recruitment, intelligence officers must evaluate the moral standards of the target's society, rather than the moral principles of a socialist country. Different cultures and legal systems have varying definitions of acceptable behavior. What may be unacceptable or punishable in a socialist society might be tolerated or even encouraged in a capitalist or non-socialist state. For example, polygamy or certain sexual behaviors that are criminalized under Soviet law may be accepted in some bourgeois societies or in traditionalist regions of the world. In some Eastern countries, particularly in Iran, members of the upper classes often maintain multiple wives—both official and unofficial—without social stigma, as this practice aligns with religious law. In certain Western states, authorities overlook or tacitly accept sexual perversions, and a thriving pornography industry contributes to the normalization of such behaviors.

Similarly, in a socialist society, theft of personal or public property in any form is morally unacceptable and incompatible with the values of socialist citizenship. In contrast, in capitalist states, particularly in the United States, large-scale financial fraud and corporate embezzlement are often perceived as bold business maneuvers rather than outright criminal acts. The moral relativity between different societies means that intelligence officers must carefully assess which moral and psychological vulnerabilities can be exploited in a given national, religious, and legal environment. The effectiveness of this recruitment approach depends on adapting to the conditions of the target country, recognizing the behavioral norms that govern the recruit's actions, and exploiting personal or social pressures that can be turned into leverage for intelligence cooperation.

A practical example illustrates recruitment on a moral and psychological basis. The agent Claude, a journalist, was assigned the task of establishing a relationship with Jeanne, an unmarried young woman serving as a secretary to an ambassador. The goal was to evaluate her suitability for intelligence work and determine whether she could be manipulated into providing classified diplomatic information.

Claude initiated contact, gradually building familiarity and trust. Over time, he established an intimate relationship with

Jeanne, frequently accompanying her to dances, restaurants, and social events. Jeanne enjoyed his company, and as their relationship deepened, she developed strong emotional dependence on him. Once Claude was confident that Jeanne had become attached to him, he subtly began to withdraw, reducing the frequency of their meetings.

Noticing the change, Jeanne reacted emotionally, questioning why he had become distant and why their meetings had become less frequent. Claude responded by feigning distress, explaining that his career as a journalist was in jeopardy. He told her that because of the time spent with her, his productivity had declined, and he now faced the risk of being fired from the newspaper. He subtly hinted that if he could maintain a steady flow of content for his editors, he would be able to spend more time with her.

Jeanne, eager to restore the intensity of their relationship, interpreted this as a request for help. Over time, she began to bring Claude classified materials that passed through her office, allowing him to copy and return them unnoticed. What had begun as an emotional dependency was transformed into a functional intelligence relationship, where Jeanne—driven by personal attachment—unknowingly became an asset providing secret diplomatic documents.

This case demonstrates how emotional vulnerabilities, personal desires, and psychological dependencies can be carefully cultivated and exploited for intelligence purposes. A well-structured recruitment operation on a moral and psychological basis does not rely on ideological commitment or financial incentives. Instead, it manipulates personal relationships, insecurities, and behavioral weaknesses to create a sense of obligation, dependency, or fear of loss, ultimately compelling the target to engage in clandestine cooperation.

Soviet intelligence occasionally recruits agents by leveraging the threat of compromise, which is considered a specialized application of the moral and psychological basis for recruitment. This method involves obtaining documents or information that, if exposed, would damage the reputation of the recruitment candidate in the eyes of their superiors, society, or family. However, the use of compromising materials is not a standard practice and is employed only when the candidate possesses exceptional intelligence value and no other recruitment basis is feasible. It is crucial that intelligence officers are certain that the target, when confronted with compromising information, will succumb to pressure and agree to cooperation rather than report the approach to authorities.

When utilizing compromising materials, it is not always necessary to apply direct pressure. Instead, it is often more effective to present the recruiter as a helpful figure, someone offering the target a way out of a difficult situation rather than an enforcer of blackmail. This approach allows for a smoother transition into cooperation, reducing the likelihood of resistance or defection.

Compromising materials typically include evidence of financial crimes, such as embezzlement of public or private funds, misappropriation of money, or corrupt dealings. They may also consist of information about personal moral failings, such as illicit affairs, sexual behavior considered unacceptable within the target's social or religious circle, or professional misconduct. Political activities can also serve as compromising material, including past or present affiliations with intelligence services of imperialist states, involvement in nationalist or anti-government organizations, or participation in plots to overthrow a capitalist regime.

However, not all compromising material is inherently useful for recruitment. It is only effective if the target is actively concealing the information from their workplace, family, or society and genuinely fears the consequences of its disclosure. If the individual is indifferent to exposure or does not regard their actions as shameful, the material loses its coercive power and is ineffective as a recruitment tool.

Before initiating a recruitment approach based on compromising materials, intelligence officers must gather thoroughly verified and well-documented evidence. General knowledge of a target's past offenses is insufficient; intelligence must possess original documents, certified copies, or irrefutable proof. If the compromising material pertains to financial misconduct, the intelligence service must have bank records, transaction logs, or internal documents confirming the misdeed. If the recruitment attempt involves a hidden extramarital affair, it is necessary to secure concrete proof, such as photographs, intercepted correspondence, or hotel records, ensuring that the material is indisputable and highly damaging if exposed.

Once compromising materials have been gathered and the target's personal psychology is well understood, intelligence officers can carefully approach the individual and strategically apply pressure. The key to success lies in framing the intelligence service as a protector rather than an extortionist, making the target believe that cooperation is their best course of action to avoid scandal, arrest, or career destruction.

Compromising materials are acquired through multiple intelligence sources. These include agents, intelligence officers stationed in both legal and illegal residencies, individuals unknowingly providing information through trusting relationships, and citizens of socialist states working in capitalist countries. Technical surveillance methods such as wiretapping, covert photography, and document interception are also instrumental in gathering incriminating evidence. However, obtaining truly damaging material is an intelligence-intensive process, requiring careful planning, operational discretion, and the use of sophisticated agent-operational techniques, as targets actively work to conceal their misconduct.

Despite its effectiveness, recruitment through blackmail produces agents who are fundamentally unreliable. Those recruited under coercion are generally hostile toward socialist intelligence services and cooperate only out of fear rather than conviction. These agents tend to evade assigned tasks, provide misleading intelligence, and attempt to minimize their involvement. Because of this, constant monitoring and reinforcement are required to ensure compliance. Intelligence officers working with such recruits must maintain strict oversight, conduct frequent checks, and ensure that the agent remains entirely dependent on the intelligence service.

To enhance control over blackmailed agents, intelligence officers must deepen their involvement in intelligence work, gradually making it impossible for them to extricate themselves from their commitments. Once an agent realizes that their continued survival, freedom, and career depend entirely on their loyalty to the intelligence service, they will become increasingly resigned to their role. At this stage, intelligence officers can incrementally shift their motivations, transitioning them from purely fear-based compliance toward material incentives, and in some cases, even ideological conditioning, depending on the specific circumstances.

Socialist intelligence services are not limited to using pre-existing compromising material. They also have the capability to exaggerate, manipulate, or even manufacture incriminating evidence when necessary. In certain cases, intelligence services employ complex operational tactics designed to trap recruitment targets in situations that make them vulnerable to blackmail. Such tactics may involve setting up financial entanglements, implicating individuals in illegal currency transactions, or luring them into compromising social or political activities that violate the laws of their home country. Once the target has been successfully maneuvered into a compromising position, intelligence officers can exploit their predicament to secure recruitment, ensuring that the agent has no alternative but to comply.

An example of a successful recruitment based on compromising materials involved a diplomat from a capitalist state. In this case, the intelligence service deliberately created a compromising situation to secure his cooperation.

A female agent was assigned to cultivate a relationship with the diplomat. Through careful maneuvering, she gained his trust to such an extent that she had access to his confidential documents and was even able to handle the keys to his safe. The intelligence service quickly capitalized on this opportunity, taking molds of the keys and producing duplicates, allowing them continued access to the diplomat's safe without his knowledge. With her assistance, all classified documents from the safe were secretly photographed, and the films were regularly passed on to the intelligence service.

After the female agent's departure, a new agent took over

the operation. While the new operative had less frequent access to the diplomat's office, he continued to remove and photograph documents from the safe. The intelligence service maintained this operation for some time, ensuring a steady flow of intelligence.

Eventually, intelligence officers identified an opportunity to recruit the diplomat himself. Knowing that he was a coward by nature, they decided to use highly classified materials—secretly extracted from his safe—to create the basis for his recruitment. The intelligence service orchestrated a controlled meeting, where the diplomat was confronted with photocopies of the missing classified documents.

During the conversation, the diplomat quickly realized that under any circumstances, the disappearance of these documents would be discovered. He understood that he would have to answer for the breach, and the consequences for losing such sensitive information could be severe, possibly even career-ending or worse. Realizing that he had no way to explain the situation and desperate to protect himself, he agreed to cooperate with the intelligence service of the socialist country.

Despite this success, recruitment based on compromising materials has not always yielded reliable agents. In several instances, individuals who were pressured into cooperation initially agreed to work with intelligence officers, only to withdraw from their commitments later. Some, once removed from the immediate threat, chose to cease cooperation and distance themselves from intelligence activities. In the worst cases, recruits turned themselves in to their country's counterintelligence agencies, exposing operations and putting both agents and networks at risk.

Intelligence officers must always consider these risks when employing compromising materials as a recruitment method. While effective in many cases, it carries significant unpredictability, particularly if the recruited individual later gains confidence, finds alternative means of escape, or chooses to report the recruitment attempt. This underscores the need for careful management, continuous monitoring, and secondary control measures to ensure that an agent recruited under duress remains compliant and does not present a security threat to intelligence operations.

Recruitment Approaches

The mere presence of a recruitment pool in capitalist countries—individuals with the necessary intelligence capabilities, personal characteristics, and a viable recruitment basis—does not automatically translate into a steady flow of agents for the intelligence services of a socialist state. The key to successful recruitment lies in the ability to effectively leverage these factors and to determine the most appropriate method for securing a target's cooperation. Selecting the right approach is essential, as it dictates how intelligence officers should proceed to ensure that a potential recruit does not reject the offer of collaboration when it is eventually made.

Not every foreigner can be approached openly or directly with a proposal for secret cooperation. A premature or improperly framed recruitment attempt risks alienating the target, alerting them to intelligence interest, or even exposing the operation to counterintelligence threats. In some cases, a more strategic and methodical approach is required, where the target is gradually brought into intelligence work without initially realizing the full extent of their involvement. However, for certain categories of recruits, a direct and immediate recruitment attempt is not only possible but advisable. The element of surprise can break their resistance, limiting their ability to refuse cooperation or seek protection from their own authorities.

This necessity gives rise to two primary forms of recruitment: gradual involvement in intelligence work and a direct proposal for secret cooperation. Each method has distinct advantages and disadvantages, and the choice between them depends on the operational environment, the intelligence value of the recruit, and, most importantly, the recruit's personal characteristics and the nature of their recruitment basis. Effective recruitment requires a careful evaluation of these factors to ensure that the chosen method aligns with the target's psychological profile, vulnerabilities, and motivations, ultimately securing their commitment to intelligence work.

Gradual Recruitment Through Progressive Involvement

The gradual involvement method of recruitment is based on the principle that an intelligence officer or agent-recruiter establishes contact with a candidate and, without revealing the true nature of their work, subtly guides them into performing tasks that align with intelligence interests. The target, believing they are assisting with personal or professional requests, gradually becomes accustomed to fulfilling assignments. Over time, they develop a trusting relationship with the intelligence officer, unaware that they have already embarked on the path of cooperation. Once the recruit has completed enough tasks and has unknowingly crossed into intelligence work, the officer tactfully discloses the true nature of their collaboration, securing a direct commitment to continued cooperation.

This method is particularly effective when recruiting on ideological, political, or material grounds. The process must be executed with patience, ensuring that the candidate receives ideological and political reinforcement while being continuously assessed and verified. Intelligence officers must be skilled in building relationships where the recruit feels an obligation to comply with their requests. Since different individuals respond to different motivations, there is no single approach to this method. Some candidates may be drawn in under the pretext of academic or literary collaboration, being asked to compile reviews, summarize statistical data, provide access to restricted but non-classified information, or share insights into government policies.

Intelligence officers in both legal and illegal residencies employ this method by initially requesting non-sensitive information, gradually increasing the complexity of the tasks assigned. As the recruit adapts to providing intelligence, they begin to understand the nature of their work but do not perceive it as a threat, believing it to be an extension of the relationship they have built with the intelligence officer. The recruiter's skill lies in leading the recruit to the conclusion that full cooperation with the intelligence service is a natural progression of their existing relationship.

This form of recruitment unfolds in two phases: a latent period, during which the recruit unknowingly contributes to intelligence operations, and an open period, in which the recruit is formally engaged and understands their role. Due to the delicate nature of this transition, the intelligence officer must exercise caution in communication. Since the recruit does not initially recognize their involvement in intelligence work, they are not equipped to take necessary security precautions. It is the intelligence officer's responsibility to ensure the safety of meetings and maintain secrecy, gradually instilling in the recruit the importance of discretion.

Secrecy is paramount, requiring intelligence officers to carefully select meeting locations based on the recruit's social, professional, and family circumstances. Ensuring that meetings remain covert and plausibly explainable is essential for the success of the recruitment process. An intelligence officer must fully understand the importance of maintaining discretion to prevent exposure.

An example of recruitment through gradual involvement on an ideological and political basis occurred when an intelligence agent in Belgium identified a promising contact, "Ler," a secretary at the Belgian embassy in a Western European country. The agent provided a tip-off about Ler, sharing a letter in which he expressed joy at the defeat of Nazi Germany and hope that Belgium would rid itself of monarchy and become a democratic republic. The residency investigated Ler's background, confirming that he came from an intellectual family, had progressive leanings, and had diplomatic service experience, making him a valuable candidate for recruitment.

An intelligence officer from a legal residency was assigned to develop Ler as a potential recruit. Knowing that Ler was passionate about water sports, the officer joined his yacht club and casually integrated into the social circle. Through carefully managed interactions, he established an acquaintance with Ler and engaged in conversations that revealed Ler's political beliefs. Ler expressed sympathy for the Communist Party and even disclosed information about British intelligence operations, including the recruitment of agents from Hungarian and Romanian prisoners in a concentration camp.

The next meeting was arranged during a fishing trip, where the intelligence officer ensured that their interaction remained discreet by avoiding direct association with Ler in the club's records. Their discussions deepened, revealing Ler's strong hostility toward British policies. Further meetings led to Ler providing confidential information about Belgian government corruption, including black-market gold sales and reactionary political dealings.

Recognizing that the time was right to escalate Ler's involvement, the intelligence officer casually requested his insights for a report on Anglo-American geopolitical contradictions. Flattered by the recognition of his expertise, Ler agreed and later submitted a detailed report compiled from Belgian business sources. At this point, the intelligence officer deemed the conditions suitable for a recruitment conversation.

Ler initially hesitated, asserting that while he was willing to support democratic causes, he did not want to formally commit to intelligence work. The intelligence officer countered by reinforcing the necessity of active engagement in the struggle for progress, arguing that Ler had already demonstrated his commitment through his past actions. Persuaded by this argument, Ler agreed to cooperate, completing his recruitment into intelligence work.

Another example of gradual involvement recruitment on a material basis involved intelligence efforts to monitor the activities of the anti-Soviet organization "Union of Ukrainians of Great Britain" (SUV) in London. The Center tasked an illegal residency with gaining access to the organization's correspondence. A district post office responsible for processing SUV's mail was identified, and a postman, "Thomson," was singled out for recruitment.

Intelligence officers conducted a detailed background check on Thomson, discovering that he was in his early fifties, had worked in the postal service for over twenty years, and was responsible for supporting his wife and two children, one of whom was disabled. His financial struggles made him a strong candidate for recruitment through material incentives. The station devised a strategy to ease Thomson into cooperation by gradually introducing him to additional income opportunities.

An experienced recruiting agent, "Harley," was assigned to handle Thomson. Using his cover as a representative of a Dutch electronics company, Harley rented an apartment within Thomson's delivery route and began receiving extensive mail. Establishing rapport with the postman, Harley casually provided small gifts and tips, subtly cultivating a relationship based on trust and financial incentives. Over time, Harley introduced new elements to the relationship, offering payment for collecting foreign postage stamps. Thomson, eager for supplemental income, willingly participated.

To further solidify Thomson's dependency, Harley created a personal scenario in which he feigned distress over suspicions about a romantic partner. Exploiting this fabricated concern, he requested that Thomson discreetly allow him to review letters addressed to the woman in question. Seeing this as a harmless favor, Thomson complied, unknowingly engaging in his first act of controlled document interception.

Once Thomson had been gradually accustomed to handling sensitive correspondence and had become reliant on the financial rewards, Harley introduced the next phase of recruitment. He proposed that Thomson provide access to mail addressed to SUV, explaining that a journalist investigating Ukrainian affairs was willing to pay generously for such materials. Thomson, now comfortable with his role and eager for additional income, accepted the offer. When he had become sufficiently enmeshed in intelligence activities, Harley formally revealed the true nature of the work, transitioning Thomson from casual assistance into structured agent cooperation with Soviet intelligence.

Harley continued political education efforts, reinforcing ideological arguments that framed Thomson's work as part of a greater cause. Eventually, Thomson became a fully committed agent, shifting from purely financial motivation to ideological conviction.

The process of recruitment through gradual involvement shows how intelligence officers transition a candidate from an ordinary relationship into a structured agent connection. The recruit is led to believe that their participation is a natural progression of their relationship with the recruiter, allowing for a seamless transition into intelligence work.

This approach presents both advantages and challenges. One advantage is that it allows for an in-depth assessment of the recruit's character, minimizing the likelihood of resistance. The recruit is gradually accustomed to intelligence tasks, eliminating the need for a sudden psychological shift. However, this method is not without risk. Not all relationships transition smoothly into intelligence cooperation, and an overeager push to formalize recruitment can backfire, leading to the breakdown of the relationship.

The greatest vulnerability of this recruitment method is the exposure risk posed by counterintelligence surveillance. In capitalist countries, enemy counterintelligence agencies closely monitor the activities and social interactions of known socialist intelligence officers operating under diplomatic or legal cover. The gradual involvement method requires repeated meetings, creating opportunities for adversary agencies to identify and track the recruitment process. If counterintelligence observes a candidate regularly meeting with an intelligence officer, they may intervene by attempting to recruit the candidate themselves, as illustrated in the French proverb, "An oak can grow from an acorn, but a pig may swallow it first."

Caution must also be exercised during early interactions. Instructing a candidate too soon about secrecy protocols can arouse suspicion and jeopardize the operation. The transition from initial contact to receiving classified intelligence may take considerable time, requiring patience and discretion. Despite these risks, the gradual involvement method remains one of the most effective recruitment strategies, allowing intelligence officers to build strong, reliable agent relationships while maintaining operational security.

Immediate Recruitment Through Direct Proposal

The direct recruitment proposal is a method in which the candidate is thoroughly studied in advance, without their knowledge, before the recruiter makes contact. The recruitment itself takes place in a single conversation, or at most, two. This method is most often used when recruiting on a material or moral-psychological basis, particularly when leveraging compromising materials. It is used less frequently for ideological-political recruitment unless circumstances strongly favor an immediate and decisive approach. There are cases when an intelligence service encounters a sudden and advantageous situation that allows for immediate recruitment without a prolonged preparatory phase.

This form of recruitment carries several advantages. The element of surprise and psychological pressure on the recruit increases the likelihood of securing their consent. It is also more difficult for enemy counterintelligence to detect interest in the recruit, as all verification and background checks are conducted covertly before the approach, meaning the intelligence officer has had no prior visible contact with the recruit. Once the recruit agrees, the intelligence officer can immediately establish the terms of their cooperation, including secret meeting procedures, information transfer methods, and financial compensation.

However, there are also considerable risks associated with this method. The recruit may refuse cooperation outright, potentially causing a scandal, reporting the recruiter to authorities, or even reacting with physical aggression. Additionally, because the recruit has not been tested in actual intelligence work, there is no certainty about their reliability or their ability to carry out assigned tasks, even if they initially agree to cooperate.

There are situations where it becomes necessary to transition from one recruitment method to another. A candidate who has been gradually involved in intelligence work may, over time, become ready for a direct recruitment conversation, making it both possible and advisable to formally offer intelligence cooperation.

A case of direct recruitment on a material basis involved the publisher of a small literary magazine, "Kon." Facing financial difficulties, he approached an intelligence agent of a socialist country, "Flynn," requesting a loan of 500 pounds sterling to pay off creditors and prevent bankruptcy. The intelligence residency initiated a study of Kon and confirmed that he had extensive connections in government circles, access to valuable political information, and was considered honest in business dealings. The investigation also verified that Kon was indeed on the brink of financial collapse.

It was decided to use his financial distress as the basis for recruitment. An intelligence officer approached Kon under the pretext of publishing an article in his magazine about the cultural achievements of a socialist country. During their conversation, Kon openly spoke about his magazine's financial struggles and the urgent need for funds. The intelligence officer, expressing sympathy, offered to lend him the required sum from government funds, but with the condition that Kon would regularly publish articles provided by the intelligence service and, in return, supply cultural and educational information. Kon readily agreed.

When the intelligence officer handed Kon the 500 pounds, he asked him to sign a receipt under the pretext that the funds were from an official source and needed proper accounting. Under the officer's dictation, Kon wrote a receipt acknowledging the loan. At this moment, the intelligence officer subtly added pressure by reminding him that the money would need to be returned by the end of the month. Kon, realizing he could not repay the debt in time, became visibly anxious. Sensing this, the intelligence officer reassured him that repayment was not necessary and that further financial assistance could be provided—on the condition that Kon discreetly supplied political intelligence. When Kon hesitated and asked whether this meant he was now a secret informant, the officer confirmed it. After some thought, Kon accepted, under the condition that their relationship remained confidential.

A case of direct recruitment on an ideological and political basis occurred with "Fellow Traveler," a man who had valuable connections, sympathized with the USSR and other socialist countries, and was openly hostile to his own government. An intelligence officer engaged in repeated discussions with him and found that Fellow Traveler's convictions were strong, and his personality was such that transitioning him into intelligence work did not require a lengthy preparation period. The recruitment was successfully completed with a straightforward proposal for cooperation, as the target had already internalized the political reasoning for his decision.

Another example of direct recruitment using a moral and psychological basis involved "Odile," a secretary-typist working at a diplomatic mission. Soviet intelligence had obtained classified materials revealing that in 1939, she had been recruited by German intelligence during World War II. Further investigation confirmed that Odile still worked at the same diplomatic mission and had recently suffered financial hardship, having lost her property during the war. She valued her current job because of its high salary and was well-trusted by her superiors. Given these factors, the residency developed a plan to recruit her using the compromising materials in their possession.

An intelligence officer initiated contact by calling Odile under the pretext of delivering a package from her brother. Intelligence had previously confirmed her brother's whereabouts to make the call appear legitimate. At the scheduled meeting, the officer apologized for forgetting the package and suggested they drive to a hotel where he had supposedly left it. Odile agreed. However, instead of heading to a hotel, the officer drove her out of town. During the journey, he revealed that he was a representative of foreign intelligence and possessed materials detailing her collaboration with the Gestapo.

At first, Odile denied everything. The officer systematically dismantled her denial by presenting undeniable details, including her Gestapo-assigned codename, the name of her German handler, and the tasks she had completed during the war. Realizing the extent of the intelligence service's knowledge, Odile broke down and admitted to her past collaboration. The officer then informed her that he was obligated to hand over these materials to the security services of her country, which maintained close relations with the socialist bloc. Facing imminent exposure and possible arrest, Odile begged for an alternative solution. The officer offered her a way out: cooperation with socialist intelligence. After a brief internal struggle, she agreed and ultimately became a valuable intelligence asset.

These examples demonstrate the effectiveness of the direct recruitment method when properly prepared. The success of these cases was due to the thorough pre-recruitment study of the candidates, ensuring that their vulnerabilities and motivations were fully understood before the recruitment attempt. The intelligence officers involved acted decisively and skillfully, using the information at their disposal to exert influence over the recruits.

Despite its advantages, direct recruitment carries significant risks. If the candidate reacts negatively, they may report the intelligence officer to local counterintelligence authorities, creating serious operational consequences. Because of this, careful study and assessment of the target is critical before proceeding with this approach. The recruit must be in a situation where they feel compelled to cooperate, whether due to ideological conviction, financial necessity, or the fear of exposure.

While the gradual involvement method allows intelligence officers time to test a recruit's reliability and capabilities, the direct recruitment proposal relies on immediate commitment, meaning the officer takes on a greater risk of failure. However, when circumstances demand urgency or when the target's situation is uniquely advantageous, the direct approach remains a powerful tool in the hands of a well-prepared intelligence operative.

Categories of Recruitment

Soviet foreign intelligence has long employed two primary types of recruitment: recruitment under one's own flag and recruitment under a foreign flag. The first requires little explanation, as it consists of recruitment conducted openly in the name of Soviet intelligence or the intelligence services of another socialist country. The second, recruitment under a foreign flag, is a more complex method that warrants closer examination.

False Flag Recruitment and Its Strategic Applications

Recruitment under a false flag involves approaching a target under the guise of another entity, typically one that the recruit sympathizes with or trusts. Given the increasing difficulty of recruitment in capitalist states due to heightened counterintelligence efforts, stricter surveillance measures, and overall hostility toward socialist nations, intelligence officers often resort to this method to secure agents who would otherwise refuse to collaborate. To accomplish this, a recruit may be approached under the pretense of working for the intelligence service of a capitalist state, a foreign diplomatic or trade mission, or a well-known political, scientific, or religious organization. Alternatively, recruitment may be conducted under the cover of large firms, banks, publishing houses, newspapers, or other institutions that hold credibility in the recruit's eyes. In some cases, entirely fictitious organizations may be created to lend further legitimacy to the recruitment effort.

The necessity of this tactic arises from the fact that key positions in the government, intelligence, and scientific sectors of capitalist states are filled with individuals who are ideologically vetted and considered politically reliable by their governments. Purges are routinely conducted within state agencies, particularly in intelligence, counterintelligence, foreign affairs, and high-security research institutions, to eliminate individuals suspected of harboring progressive or leftist sympathies. As a result, many of the most strategically placed individuals are fundamentally hostile to the socialist bloc, making direct recruitment nearly impossible. The use of recruitment under a false flag allows intelligence officers to bypass ideological resistance by appealing to recruits under the banner of institutions they already respect or trust.

Conducting recruitment under a false flag requires exceptional skill and discretion. When targeting individuals who are politically aligned with capitalist governments, intelligence officers must take great care to ensure that no aspect of the recruitment process inadvertently reinforces the recruit's bourgeois ideology or strengthens their loyalty to imperialist states. Intelligence officers operating from both legal and illegal positions have the ability to conduct false-flag recruitments personally or through intermediaries. Illegal intelligence officers, in particular, have greater freedom to assume alternative identities and may present themselves as representatives of third-country intelligence or counterintelligence services. They can also use cover as business executives, financial consultants, journalists, religious officials, or members of international industrial or professional associations, depending on what is most appropriate for the target.

Legal residencies, while offering certain advantages, come with the limitation that intelligence officers are often publicly recognized as representatives of socialist states, making it more difficult for them to assume a foreign flag directly. However, they can still conduct false-flag recruitments through intermediary agents with the same level of effectiveness as their illegal counterparts. In either case, the recruitment process must be executed with the highest level of vigilance and secrecy. Not only must the recruit's cooperation be secured without alerting local counterintelligence agencies, but the recruit must also remain unaware of the recruiter's true identity and their connection to socialist intelligence.

Those conducting recruitment under a false flag may include intelligence officers from the Center, citizens of socialist countries traveling in capitalist nations, or long-term residents abroad who serve as undercover operatives. Additionally, carefully vetted agents from third-party countries may be used as intermediaries, allowing the intelligence service to maintain deniability and further obscure its true involvement in the recruitment process.

Criteria for Selecting Candidates for False Flag Recruitment

The expansionist ambitions of the leading imperialist powers and the economic and political aspirations of capitalist monopolies create continuous conflicts across various regions, including the Far, Middle, and Near East, Africa, and Latin America. These struggles also manifest within the imperialist states themselves, drawing in various social strata with competing interests. In any given capitalist country, factions of the national bourgeoisie are often engaged in internal struggles, dividing themselves into opposing groups that align with different foreign powers to further their own economic and political agendas. These conflicts extend beyond governments to include monopolies, banks, political parties, press organizations, religious groups, and other bourgeois institutions, all vying for dominance.

Anti-American sentiment is particularly strong in Latin America, as well as in many Asian, African, and European countries. Anti-British and anti-French sentiments have also been growing, particularly in former colonial territories where national liberation movements continue to resist the economic and political dominance of Western imperialist powers. Parts of the national bourgeoisie, both petty and large, are increasingly drawn into these movements as they seek to protect their interests from foreign monopolies. In countries such as Pakistan, where British and American imperialist interests clash, elements within the local bourgeoisie take opposing sides, aligning themselves with either British or American capitalists. A similar situation exists in Iran, Saudi Arabia, and other regions where imperialist competition fuels internal divisions.

Given these conditions, intelligence officers from socialist states operating in both legal and illegal residencies have ample opportunities to identify and recruit individuals across various strata of society. This can be done under the guise of representing English, American, French, or other Western intelligence services, as well as monopolies, bourgeois political parties, religious organizations, and other institutions that the candidate for recruitment sympathizes with. In certain cases, intelligence officers can even infiltrate bourgeois parties or factions competing for power within capitalist states. By positioning themselves as representatives of organizations that align with the recruit's political or economic interests, socialist intelligence officers can secure cooperation while maintaining operational cover.

The decision to recruit under a false flag depends on whether such an approach will effectively fulfill the intelligence service's objectives and whether the chosen cover is the most suitable for securing the recruit's trust. Recruitment under the flag of a capitalist intelligence service is often more advantageous than recruitment under the cover of a political party or other organization. A recruit who believes they are working for a Western intelligence service is more likely to accept direct intelligence tasks, follow secure operational protocols, and develop conspiratorial discipline, making their work more productive and reducing the risk of exposure.

However, some candidates may refuse to collaborate with intelligence services outright, regardless of the cover story. In such cases, it may be more effective to recruit them under the guise of a political, economic, or religious organization. There have been instances where individuals who categorically rejected offers from so-called "bourgeois intelligence agencies" nevertheless agreed to perform tasks for political or economic organizations they believed were aligned with their interests.

By applying Marxist-Leninist analysis to the class structures of capitalist societies, intelligence officers can effectively identify and recruit individuals under the flag of various Western institutions, whether state-controlled or private. The process of identifying potential recruits follows the same fundamental principles as recruitment under a true socialist flag. Leads are obtained through the study and assessment of target institutions, and the selection process is carried out with the broader goals of the intelligence service in mind. Sources of recruitment leads remain the same, including intelligence agents and other informants.

The study and verification of potential recruits follow the same rigorous procedures as when recruiting under a socialist flag. However, one unique challenge of false-flag recruitment is ensuring that the intelligence officers and agents involved never unintentionally strengthen the recruit's reactionary or counterrevolutionary views. Agents tasked with verifying leads must engage with their targets in a manner consistent with the false flag being used while avoiding any action that could reinforce anti-socialist ideologies.

When intelligence officers from legal and illegal residencies study a candidate for recruitment under a false flag, they must ensure that the candidate has no prior knowledge of them as representatives of a socialist state. They must also demonstrate complete familiarity with the identity they assume, including fluency in the appropriate language, knowledge of relevant cultural and historical facts, and possession of authentic-looking documentation supporting their cover. If a recruiter poses as an American intelligence officer, they must speak English fluently with an appropriate accent, possess a deep understanding of American society and politics, and carry credentials that support their assumed identity. If posing as a French national recruiting for American intelligence, it is not necessary to have in-depth knowledge of American affairs, but they must be entirely credible as a French citizen.

Another critical factor is ensuring that the recruit is not, in fact, associated with the intelligence or counterintelligence service under whose flag the recruiter is operating. A failure to confirm this could lead to an adversarial counterintelligence officer being unwittingly recruited, exposing the operation and compromising the intelligence network.

The selection process involves eliminating unsuitable candidates while taking the most promising leads into active recruitment development. Intelligence officers move forward only with individuals who demonstrate the ability to carry out intelligence tasks and whose motivations, character, and behavioral tendencies indicate they will respond positively to recruitment under a false flag.

Psychological and Strategic Justifications for False Flag Recruitment

When agents are recruited on an ideological and political basis under the true flag of a socialist intelligence service or another organization from a socialist country, the process relies on a shared ideological worldview between the recruiter and the recruit. In these cases, the recruit agrees to cooperation because they believe their political beliefs align with those of the recruiter, and they see their collaboration as part of a greater ideological struggle. The foundation of such recruitment is the mutual recognition of political values, which provides a strong and often stable motivation for the recruit's commitment.

In contrast, when recruitment is conducted under a false flag, no such ideological commonality exists between the recruiter and the recruit. The recruit mistakenly believes that the recruiter represents an organization or intelligence service that aligns with their own political beliefs, personal inclinations, or professional interests. This deception is fundamental to false-flag recruitment. The recruit consents to cooperation because they assume that the recruiter holds the same values or national loyalties as they do, leading them to believe they are assisting a capitalist intelligence agency, a respected political organization, or an entity they trust.

Material incentives and compromising materials can also be used in false-flag recruitment. The choice of which flag to use is determined by the political preferences of the recruit, making it easier to gain their cooperation. A recruiter must always operate under a flag that aligns with the recruit's existing loyalties. If an individual is known to sympathize with the British, attempting to recruit them under the flag of French intelligence would be counterproductive. Instead, the recruitment should be conducted under the British flag, as the recruit's pro-British leanings will serve as an additional motivation for cooperation. By carefully selecting the most suitable false flag, recruiters increase the chances of a successful recruitment operation while minimizing resistance from the recruit.

There are also special circumstances where false-flag recruitment is dictated by the nature of the mission. In cases involving sabotage, terrorism, or high-risk covert operations, it would be extremely dangerous and counterproductive to reveal that a socialist intelligence service is directing the recruit's activities. Operating under a false flag in such situations protects the socialist state from potential blowback, ensuring that if the recruit is exposed or the operation is compromised, responsibility is deflected onto a bourgeois intelligence service. This method significantly reduces the risks associated with active measures, making it a crucial tactic for covert operations that require plausible deniability. Agents involved in such missions are almost always recruited under the false flag of a capitalist intelligence or counterintelligence service and motivated primarily by financial incentives, rather than ideological alignment.

Execution and Finalization of False Flag Recruitment Operations

The recruitment process under a false flag begins with an extensive study of the target. This phase is critical in determining the basis for recruitment, assessing the intelligence value of the candidate, and identifying the most effective approach. Once this preparatory work is completed and it is confirmed that the target can be recruited under the chosen false flag, intelligence officers must carefully select the method of recruitment and the recruiter best suited for the task.

False-flag recruitment follows the same two fundamental approaches as recruitment under one's own flag: gradual involvement or direct proposal. The choice of method depends on multiple factors, including the recruit's motivations, personality, and the intelligence environment of the country where the operation takes place. The gradual involvement method is often preferable when dealing with individuals who are ideologically aligned with bourgeois or imperialist interests. This approach allows intelligence officers or agent recruiters to build personal relationships, gain trust, and progressively introduce the recruit to intelligence tasks. However, when the recruit has been sufficiently studied and there is confidence that they will accept cooperation without prolonged engagement, a direct proposal may be used instead. This is particularly useful when operational circumstances require a swift recruitment, such as when either the recruit or the recruiter has only a short period in the country. Even in cases where a direct proposal is used, it must be preceded by a detailed and thorough study of the recruit.

Gradual involvement is also frequently used when recruiting under a false flag on a material basis. In contrast, direct recruitment is the preferred approach when recruitment is based on moral-psychological factors, particularly the use of compromising materials. In such cases, the recruit is already vulnerable to exposure, and intelligence officers can leverage this pressure to secure immediate cooperation. Regardless of the chosen method, recruiters must carefully assess the operational context and the personal characteristics of the recruit to ensure the most effective approach. In some cases, it may be more effective to use direct recruitment rather than gradual involvement, even when dealing with a recruit who holds strong bourgeois views. Similarly, a candidate who might traditionally be recruited through a direct proposal due to financial vulnerability may instead be drawn into intelligence work more gradually.

A well-developed cover story is crucial in false-flag recruitment. The recruiter must fully understand the fabricated identity they are assuming and be able to convincingly maintain it. Even the slightest inconsistency in their legend can jeopardize the entire operation and expose the intelligence service's true involvement.

An example of successful false-flag recruitment occurred in 1935 in Austria. Soviet intelligence received information on "Janus," an official in the Austrian Ministry of Foreign Affairs who had access to highly classified documents. Janus was known to be a strong supporter of Austria's annexation by Germany and had ties to the National Socialist movement. Given these political leanings, it was decided to recruit him under the flag of German intelligence. The recruitment operation was assigned to "Winkel," a trusted agent who was an Austrian national and worked in the Austrian branch of a major German trading firm. Winkel's brother had been arrested by Austrian authorities for his role in the failed Nazi coup of July 1934.

Through surveillance, it was established that Janus frequently visited the Mozart Café after work. Winkel began frequenting the café at the same time, gradually making himself familiar to Janus. One day, Winkel initiated a casual discussion about an article in the newspaper, and from that point, their conversations became more regular. As their relationship developed, Winkel confirmed that Janus was not only committed to German fascism but also struggling financially. When Janus learned about Winkel's brother's arrest, he became more sympathetic toward him, strengthening their bond.

Over time, Winkel guided their conversations toward the necessity of action in support of National Socialism. Janus responded that while he believed in the cause, he was reluctant to join the underground Austrian Nazi movement because of its lack of discipline and security. He feared that his involvement would be discovered and that his career in the Ministry of Foreign Affairs would be ruined. Winkel assured him that he had something more serious in mind—a top-secret, well-organized, and well-financed operation that could bring real results. He then proposed that Janus collaborate with German intelligence. Janus accepted the offer and went on to provide critical intelligence on Austrian foreign policy for several years.

In modern intelligence work, the role of recruiters operating from illegal positions has become increasingly important, as has the role of agent recruiters who are capable of recruiting under a false flag.

Another example of false-flag recruitment involved a German agent within the Soviet intelligence network who had long-standing ties to anti-Soviet White Guard organizations. Among his acquaintances was a prominent White Guard leader who regarded the agent as a representative of German industrial and political circles, believing he was an informal contact for the German embassy. Soviet intelligence used this misconception to its advantage, recruiting the White Guard leader under the flag of German intelligence. The recruit spent years working for Soviet intelligence, unaware that he was serving the very regime he sought to undermine.

A similar case involved a Soviet intelligence residency that had an agent named "Dombey" in its network. The residency discovered that Dombey maintained a friendly relationship with "Boss," a senator who was a leading figure in the right-wing opposition. Further study revealed that Boss was under significant financial strain, supporting a large family on a modest senator's salary. His attempts to supplement his income through journalism were proving inadequate. Given his financial difficulties, the residency decided to recruit him under a false flag.

Since Dombey had previously worked as a journalist and had strong connections to British media, it was decided that Boss would be approached under the flag of the right wing of the British Labour Party. Dombey proposed to Boss that he provide confidential political intelligence to key Labour leaders in exchange for financial compensation. Because of Dombey's well-known British connections, Boss saw nothing unusual in the request and agreed. He subsequently passed on sensitive political information, believing he was strengthening ties with British political allies, when in reality, he was working for Soviet intelligence.

Once an agent is recruited under a false flag, they typically remain in contact with their original recruiter. However, if circumstances require a transfer to another intelligence officer, it is crucial that the new handler fully aligns with the false-flag identity. Agents recruited under false flags must never suspect they are being handed over to a different intelligence service. If a recruiter must leave the country or return to their permanent position, they must ensure that the transition is seamless.

The timing and location of a recruitment conversation under a false flag is of critical importance. A recruitment meeting must never be held in places where other agents are received, nor in locations where there is any risk of exposing the recruiter or revealing the operation. Strict operational security measures must be in place to prevent counterintelligence detection. Every aspect of the recruitment—from the selection of the candidate and the choice of flag to the structuring of future interactions—must be meticulously planned to ensure success and minimize risk.

Transitioning a Recruited Agent from a False Flag Cover to Direct Cooperation with Socialist Intelligence

Recruiting agents under a false flag presents limitations, particularly in terms of the intelligence tasks they can be assigned. An agent recruited under the pretense of working for a capitalist intelligence service cannot be expected to gather information that would not align with the supposed interests of their recruiter. They cannot be tasked with collecting intelligence for a socialist state without their eventual realization that they have been deceived. To fully utilize the agent's capabilities, intelligence officers must gradually entangle them in operations, making them dependent on the intelligence service and, when possible, transitioning them into direct cooperation under the flag of a socialist country.

The transition from false-flag recruitment to direct recruitment under a socialist flag is a delicate process that requires time and extreme caution. It cannot be rushed, as premature exposure of the agent to the true nature of their handlers may lead to their defection or outright betrayal. The case of "Fairy" demonstrates the successful application of this transition. A secretary-typist at an American diplomatic mission, Fairy was known to harbor anti-American sentiments and did not consider the United States her homeland, having been taken there as a child by her emigrant parents. Recognizing this ideological vulnerability, the residency deployed an agent, Robert, a citizen of the country Fairy still considered home, to approach her.

Through careful psychological maneuvering, Robert cultivated a relationship with Fairy, playing on her personal sympathies and nationalist sentiments. He acted as if he was working for his country's intelligence service rather than for a socialist state. Over time, Fairy agreed to pass on secret information about the Americans, beginning with verbal reports and eventually escalating to the transfer of classified documents from the American embassy. Once she had engaged in multiple intelligence operations and was sufficiently committed, Robert gradually conditioned her to adopt stronger anti-American views, reinforcing her alienation from the country she worked for.

When the time was deemed right, the transition to open cooperation with the socialist intelligence service was carefully staged. A dinner was arranged where Fairy was introduced to a residency officer who had been overseeing her case. During the conversation, the officer carefully revealed that the authorities of his country were aware of her work with Robert. He explained that he had refrained from telling her earlier but now believed she deserved to know the truth, as she had already exposed herself to risk. He assured her that by openly committing to the intelligence service of a socialist state, she would be able to fight American imperialism more effectively. Initially taken aback, Fairy quickly overcame her surprise and affirmed her willingness to continue working against the Americans.

The process of ideological and political reinforcement plays a crucial role in determining the success of transferring an agent to work under a socialist flag. The prevailing political climate in the recruit's country, their personal disposition, and the extent of their dependence on intelligence operations all influence their decision. If an agent remains hostile to socialism, any attempt at direct ideological influence risks exposing the recruiter's true allegiance. In such cases, the transition must be handled with the utmost caution, ensuring that the agent remains unaware of the deception until their position is so compromised that defection is no longer an option.

There have been instances in which even recruits with strong anti-socialist sentiments were successfully transferred under the right circumstances. During World War II, a high-ranking Polish officer in Anders' Army was recruited under the guise of working for British intelligence. Convinced that he was serving British interests, he provided valuable intelligence to Soviet handlers for an extended period. As the tide of war shifted in favor of the Soviet Army, it became increasingly important to repurpose him for operations against the British and Americans. The obstacle was his belief that he was working for British intelligence. Given the changing military and political situation, Soviet intelligence determined that the time was right to reveal the truth.

The officer was stunned when he learned that all the intelligence he had provided had been sent to the Soviets rather than the British. However, by the time of this revelation, his situation had changed. His intelligence work had been extensive, and breaking ties with the Soviets was not an option without severe consequences. He chose to continue providing intelligence, now knowingly working for Soviet intelligence, despite his initial opposition. This case demonstrates that the success of such a transition depends not only on operational precision but also on a correct assessment of the recruit's psychological state and external circumstances.

In the modern intelligence landscape, false-flag recruitment remains an invaluable tool for socialist intelligence services. Intelligence officers operating from both legal and illegal residencies should not limit themselves to seeking agents willing to cooperate directly with socialist countries. By utilizing false flags strategically, they can expand their recruitment opportunities in capitalist states, penetrating deeper into enemy institutions. Agents recruited under a false flag widen the operational scope of socialist intelligence, creating additional avenues for gathering crucial intelligence and carrying out sensitive operations.

Psychological and Tactical Methods of Persuasion in Recruitment

At first glance, the means available to an intelligence officer to influence a recruit may seem limited, as they boil down to two fundamental approaches: persuasion and coercion, or a combination of both. However, within these broad categories, there exist numerous variations, each tailored to the psychological profile, vulnerabilities, and circumstances of the recruit. Persuasion can take many forms, including direct conversation, appeals to logic or emotion, demonstrations of ideological alignment, references to the recruit's personal experiences, historical examples, literature, films, and even personal relationships. An intelligence officer can build trust, establish credibility, and create a sense of obligation in the recruit through gradual psychological conditioning, reinforcing the idea that cooperation is a natural and beneficial choice.

Coercion, on the other hand, operates through pressure and the strategic application of fear. Economic constraints can be exploited to push a recruit into compliance, while legal or political threats—such as exposing illegal activities, security breaches, or personal indiscretions—can be leveraged to force cooperation. Public opinion, social standing, and family influence can also be used as tools of coercion, as a recruit may fear disgrace or professional ruin more than any direct threat. In extreme cases, the mere suggestion of physical danger can serve as a powerful motivator, though this method is rarely used by intelligence services of socialist states.

An intelligence officer has at their disposal a wide array of methods to exert intellectual, emotional, and psychological influence over a recruit. The recruit often perceives the intelligence officer as a figure of authority, intelligence, or charisma, or as someone who combines all these attributes. The officer must assess which aspects of persuasion will be most effective—whether appealing to the recruit's personal ambitions, ideological leanings, financial pressures, or emotional dependencies.

In the vast majority of cases, socialist intelligence agencies rely primarily on persuasion to secure cooperation, carefully cultivating a recruit's willingness through psychological engagement and subtle reinforcement. Coercion is only used in exceptional circumstances, where no other option exists, and even then, it is applied in a controlled and measured manner to ensure the recruit remains operationally useful and does not become a liability.

PART IV

DEVELOPMENT, EXECUTION, AND FINALIZATION OF RECRUITMENT OPERATIONS

Defining the Recruitment Development Process

Recruitment development refers to a systematic set of intelligence measures designed to covertly assess a potential recruit, evaluate their intelligence capabilities and personal characteristics, determine the most effective recruitment approach, and create the necessary conditions for a successful recruitment operation. It is a preparatory stage in which intelligence officers gather detailed information on the target to ensure that the recruitment process is executed with precision and minimal risk.

Operational Objectives of the Recruitment Development Phase

Several key tasks must be completed during the development phase to ensure a successful recruitment. A thorough examination of the individual is conducted to refine and confirm the most effective recruitment basis. Intelligence officers work to clarify the recruit's access to valuable information, evaluating how their position and connections can serve the needs of the intelligence service. Simultaneously, personal qualities such as reliability, discretion, and psychological resilience are carefully assessed.

Once these aspects are understood, intelligence officers determine the appropriate form of recruitment, whether it will follow a gradual involvement approach or a direct proposal, and whether recruitment will take place under a true or false flag. The selection of a recruitment method is tailored to the specific vulnerabilities, motivations, and personality traits of the candidate. The intelligence service also identifies the best method of approaching the recruit, which includes choosing whether the recruitment should be carried out by a professional intelligence officer or through an intermediary agent who may have a more natural connection to the target.

Each stage of recruitment development is designed to ensure that by the time the recruit is approached, the intelligence service has maximized its chances of success while minimizing the risk of exposure or failure. The process is not rushed, as a premature or poorly planned recruitment attempt can not only lead to rejection but also jeopardize future operations. Through careful study and strategic planning, intelligence officers create an environment in which the recruit is most likely to accept cooperation willingly or, in cases where necessary, with limited resistance.

Refining the Recruitment Basis

After initiating a recruitment operation, intelligence services continue their in-depth study of the selected individual through agents and other means. This ongoing investigation aims to refine the chosen recruitment basis, ensuring that the initial assessment was accurate. At the same time, intelligence officers search for additional factors in the target's worldview, daily life, and financial situation that could serve as supplementary motives to facilitate recruitment. The process of refining the recruitment basis is essential for increasing the likelihood of success while minimizing resistance from the recruit.

When developing individuals whose recruitment is planned on an ideological and political basis, intelligence officers focus on understanding the depth of their political views and beliefs. They analyze the candidate's stance on key political and economic issues, identifying areas where their views align, either fully or partially, with those of the socialist intelligence service. From these areas of ideological alignment, the recruiter must select the most compelling issues that could serve as the primary motivation for recruitment. Efforts are then made to reinforce the chosen ideological foundation, strengthening the target's existing beliefs to ensure their commitment to the cause. Intelligence officers also prepare carefully reasoned arguments designed to persuade the candidate to agree to secret cooperation, framing their participation as a necessary and logical extension of their existing convictions.

The experience of socialist intelligence services demonstrates that successful ideological and political recruitment requires active engagement to refine the recruitment basis. It is not enough to identify a recruit's general ideological alignment; intelligence officers must pinpoint the specific elements that will be most persuasive in securing their cooperation. Every aspect of the recruit's political outlook must be examined and, where possible, reinforced to ensure their long-term commitment.

When specifying the material basis for recruitment, intelligence officers work to gain a deeper understanding of the recruit's financial situation. They identify the key financial pressures affecting the recruit and assess which factors are most critical to their well-being. Intelligence must determine what financial incentives would be most effective—whether offering a profitable business opportunity, direct monetary compensation, or valuable gifts. Once the recruit's financial weaknesses are fully understood, intelligence services take measures to reinforce the chosen recruitment basis, ensuring that the target becomes dependent on the intelligence service as their primary source of financial relief.

In cases where recruitment is based on compromising materials, intelligence officers seek to verify and strengthen the incriminating information. This involves confirming the exact circumstances under which the recruit engaged in compromising activities and ensuring that these actions, if exposed, would pose the greatest threat to their reputation, career, or personal life. The intelligence service documents these actions meticulously, creating an undeniable body of evidence. Simultaneously, officers identify any auxiliary motives that may further incentivize the recruit to cooperate, such as professional ambitions, social status, or personal security concerns. To ensure maximum leverage, steps are taken to reinforce the incriminating materials, securing additional documentation or witness testimony that makes the recruit fully aware that their only viable course of action is to cooperate with the intelligence service.

Confirming Intelligence Value and Operational Capabilities of the Recruit

Simultaneously with refining the recruitment basis, intelligence officers must thoroughly reassess the target's intelligence capabilities, even if they initially appeared indisputable at the onset of development. The process of verification may uncover new, previously unidentified opportunities that enhance the recruit's value to intelligence operations. However, it may also reveal the opposite—that the individual lacks the access or influence originally attributed to them or has lost these capabilities over time due to changes in their professional position or personal circumstances.

Clarifying intelligence capabilities ensures that each recruit is assigned tasks that align with their actual access and influence. A recruit with verified and well-documented intelligence potential can be effectively positioned within operations, maximizing their contribution to intelligence gathering. Conversely, if the study reveals that the recruit's capabilities are limited or have diminished, intelligence officers must reassess whether further efforts are justified. It may be determined that the development process should be terminated to avoid wasting time, resources, and operational efforts on a recruit who no longer holds strategic value.

The ultimate objective of this verification process is to obtain a deeper and more precise understanding of the recruit's intelligence potential. By doing so, intelligence officers can determine the best possible use for the recruit within intelligence work or, if necessary, abandon further development in favor of more promising candidates. The refinement of intelligence capabilities serves as a critical checkpoint, ensuring that recruitment operations remain efficient, targeted, and aligned with strategic intelligence goals.

Detailed Psychological Profiling of the Recruitment Target

When conducting recruitment development, a thorough

understanding of the personal qualities of the individual being developed is of critical importance. As the recruitment process nears its decisive phase, intelligence officers must have a precise assessment of both the psychological and physical characteristics of the recruit. These attributes influence not only the choice of recruitment form but also the specific approach that will be taken during the recruitment conversation itself. A miscalculation in evaluating the recruit's temperament, behavioral tendencies, or emotional resilience can result in rejection, failure, or even exposure of the operation.

A comprehensive identification of the recruit's personal qualities is conducted alongside the refinement of their recruitment basis and intelligence capabilities. This is achieved through intelligence agents and other covert means, ensuring that no crucial detail is overlooked. In practice, there are frequent cases in which new personality traits emerge during development that significantly impact the recruitment process. Such discoveries may necessitate adjustments in timing, modifications to the recruitment strategy, or in some instances, a decision to abandon the recruitment effort altogether.

For example, if a candidate is excessively talkative, the gradual involvement method must be adapted accordingly. Special care must be taken in selecting discreet locations for meetings so that the recruit does not inadvertently expose their relationship with intelligence officers. Additionally, ongoing engagement must subtly reinforce the importance of secrecy, ensuring that the recruit internalizes the necessity of maintaining absolute discretion in their dealings with a representative of a socialist intelligence service. Without this precaution, the recruit may unwittingly disclose sensitive information, placing both themselves and the intelligence operation at risk.

In cases where a candidate possesses a strong-willed personality and is likely to resist recruitment despite the use of compromising materials, additional precautions must be taken to ensure the intelligence officer or agent-recruiter maintains control of the situation. The setting for the recruitment conversation must be carefully chosen to prevent the recruit from easily leaving or abruptly terminating the discussion. Measures should also be in place to capture the interaction, such as recording the conversation to reinforce the gravity of the situation. The intelligence officer must structure the meeting in a way that allows ample time for persuasion, ensuring that any initial resistance is systematically dismantled through a combination of psychological pressure, rational argumentation, and strategic maneuvering. The success of the recruitment conversation depends on creating conditions in which the recruit feels compelled, whether by conviction or necessity, to accept cooperation.

Selection of an Appropriate Recruitment Approach

Simultaneously with refining the recruitment basis, verifying intelligence capabilities, and thoroughly assessing the personal qualities of the candidate, intelligence services must carefully determine the appropriate recruitment form. The choice is between a direct offer of cooperation or gradual involvement in intelligence work. The decision depends on several factors, including the recruitment basis, the personality and psychological disposition of the candidate, the broader intelligence and operational environment, and the recruiter's own characteristics and abilities.

When recruitment is based on ideological and political alignment, the candidate agrees to cooperate due to shared political beliefs and convictions. Progressive, democratically minded individuals often see their involvement as part of a greater struggle for their own nation's interests. However, a key challenge in recruiting such individuals is persuading them that working with a socialist intelligence service is a continuation of their existing struggle, albeit through different methods. Many politically motivated recruits initially hold the belief that the fight for peace and justice should be waged openly rather than through clandestine means. The intelligence officer's task is to gradually reshape this perception, convincing them that covert intelligence work is, in fact, the most effective way to achieve their ideals. Because of this psychological adjustment, intelligence officers typically introduce such recruits to intelligence work in stages, ensuring they become comfortable with small tasks before fully committing to secret cooperation.

However, if the recruit has been extensively studied and there is confidence that they will cooperate without the need for gradual engagement, a direct proposal can be used. In such cases, the intelligence officer presents the candidate with a straightforward offer of collaboration, appealing to their ideological convictions without requiring preliminary conditioning.

The broader intelligence and operational climate in a given country also plays a crucial role in determining the recruitment approach. In many capitalist states, particularly in the United States and its allied countries, counterintelligence and law enforcement agencies impose restrictions on contact between local citizens and officials from socialist states. These restrictions include enhanced surveillance, controlled travel, and extensive propaganda campaigns against socialist countries, creating a hostile environment for intelligence work. In such situations, prolonged recruitment efforts pose a higher risk of exposure. Therefore, intelligence services may need to accelerate the recruitment process, relying on direct recruitment proposals rather than extended development phases. Candidates can be deeply studied through agents, with the intelligence officer only stepping in at the final stage to conduct recruitment, thereby minimizing face-to-face contact and reducing operational risks.

The choice of recruitment form must always be approached with flexibility rather than rigid adherence to standard procedures. Intelligence officers often encounter cases that do not conform to established guidelines and must adapt based on their own experience and the specific conditions of each case. Creative problem-solving is essential to determining the most effective method for recruitment.

For example, an operative received a tip about an individual known as "Verdi", an artist with significant social and political connections among officials in his country's state apparatus. The report indicated that Verdi was dissatisfied with his government's pro-American policies and frequently expressed democratic, progressive views. This suggested that Verdi might be willing to cooperate with a socialist intelligence service due to his ideological leanings. However, Verdi's visit to the country was brief, lasting only a few days, which ruled out the possibility of a gradual recruitment approach.

Considering the time constraints, the residency consulted with the Center and decided to proceed with a direct recruitment proposal. To ensure operational security, they selected an intelligence officer who was scheduled for recall to his home country, minimizing potential risks. If Verdi refused the offer, the intelligence officer could depart without jeopardizing the residency's overall work.

The recruitment conversation was successful, confirming that the choice of a direct approach had been the correct one in Verdi's case. His short stay in the country was the decisive factor in the recruitment strategy, demonstrating that intelligence operations must always be adaptable to the specific circumstances of the recruit and the surrounding operational environment.

Designating a Recruiter and Integrating Them into the Recruitment Process

During the process of recruitment development, the issue of selecting the appropriate recruiter to handle the recruitment of the target individual arises. The recruiter may be an intelligence officer from either a legal or illegal residency, an operational worker from the central intelligence apparatus, a recruited citizen of a socialist country stationed in a capitalist state, or a trusted agent from among the citizens of a bourgeois country. Each of these categories presents different advantages and operational considerations.

In the increasingly complex intelligence environment faced by socialist intelligence services, the role of agents as recruiters has grown significantly. Counterintelligence and law enforcement agencies of hostile states impose numerous obstacles on recruitment activities, particularly against the legal intelligence apparatus. One way to circumvent these obstacles is by relying on loyal and experienced agent-recruiters who already have access to potential recruits and can conduct recruitment efforts with lower risk of exposure.

The decision on which recruiter to use and how to approach the candidate is typically made during the study and verification of leads. Intelligence officers conducting these investigations develop a deep understanding of the recruit's strengths and weaknesses, sometimes even exerting influence over them during the development phase. This familiarity often determines the selection of the recruiter. In many cases, agents themselves provide tips on potential recruits from within their personal networks—friends, colleagues, acquaintances, or relatives—and become directly involved in their development. Because these agents already maintain established relationships with the targets, they are well aware of their political sentiments, personal needs, vulnerabilities, and ambitions, making it easier for them to facilitate recruitment. This natural rapport reduces suspicion and increases the likelihood of success, a factor that is always taken into account when selecting a recruiter.

Using an agent-recruiter who already has a relationship with the target offers additional advantages in terms of operational security. In such cases, only the agent-recruiter's identity is revealed to the recruit, maintaining a high level of conspiracy. Additionally, no elaborate measures are needed to create a plausible approach to the recruit, as the recruiter is already in regular contact with them.

In other cases, it may be necessary to assign the recruitment task to intelligence officers or agents who have had no prior direct contact with the recruit. These recruiters are introduced into the recruitment process either soon after the development phase begins or at its final stage, when recruitment is ready to proceed. This is typically necessary when an intelligence officer or agent involved in the recruit's development must be withdrawn from the operation. Such withdrawals may be required for security reasons, such as diverting counterintelligence attention in the event of a recruitment failure. In other instances, the intelligence officer or agent originally involved in development may not be suitable for the recruiter's role due to their personal characteristics, professional background, or social standing.

When a new recruiter must be introduced to the recruit,

this process follows the same careful approach used during the development, verification, and study of leads. The recruiter must establish personal contact with the recruit without an intermediary, using a plausible pretext that aligns with the recruit's personal and professional circumstances. The initial meeting may be arranged under the guise of official business, personal connections, social engagements, or professional networking. Encounters at receptions, cultural or sporting events, academic meetings, resorts, clubs, or restaurants serve as common settings for these introductions. A letter of recommendation from a mutual acquaintance can also be used to facilitate contact between the recruiter and the recruit in a natural and credible manner.

The method chosen to establish contact with a recruit plays a critical role in the success of the recruitment effort. A poorly conceived or clumsily executed approach may immediately arouse suspicion, causing the recruit to distance themselves or reject further interaction. A misstep in the initial contact can also draw the attention of counterintelligence agencies, jeopardizing not only the recruitment attempt but potentially exposing the intelligence operation itself. Ensuring that the introduction appears natural and free of suspicion is therefore essential for maintaining secrecy and achieving a successful recruitment outcome.

Strategic Timing and Location Selection for the Recruitment Conversation

In bringing recruitment development to completion, intelligence must carefully determine the most opportune time and setting for the recruitment conversation, as these factors can significantly influence the outcome. The timing of the conversation must be chosen strategically to align with the recruit's psychological and material state, ensuring that they are most receptive to the proposal.

When recruiting a candidate on a material basis, the recruitment conversation should be conducted at a moment when the recruit is experiencing heightened financial distress and is actively seeking a way out of their difficulties. This maximizes the effectiveness of the material incentive and increases the likelihood that the recruit will agree to cooperation.

For recruitment based on ideological and political alignment, where the recruit has typically been gradually drawn into intelligence work, the conversation must take place at a point where the recruit has developed sufficient trust and ideological commitment to make continued indirect engagement unnecessary. At this stage, the recruit must be persuaded that formal cooperation with intelligence is the logical next step in their commitment to their political beliefs.

When recruiting on a moral and psychological basis using compromising materials, the timing of the recruitment conversation is particularly critical. The conversation must be held when the compromising materials are at their most effective, meaning when the recruit is at peak vulnerability and most fearful of exposure. A premature approach could allow the recruit to seek protection elsewhere, while a delayed approach risks the compromising information becoming less useful due to changes in the recruit's circumstances.

Selecting an appropriate location for the recruitment conversation is equally important. The setting must allow the recruiter to maintain control over the situation and ensure security in the event of a failed recruitment attempt. Experience in capitalist countries has shown that suitable locations for recruitment conversations include recreational areas, resorts, cafés, and restaurants, where the recruit is likely to feel at ease and the conversation can proceed naturally.

Conducting the conversation in outdoor locations such as beaches, countryside retreats, or fishing spots offers the advantage of privacy and allows for a calm and controlled discussion. In some cases, arranging the recruitment conversation in a different city or even another country under the pretext of a leisure trip can further reduce the risk of exposure.

Resorts and tourist centers offer additional advantages, as they provide a setting where people from different regions and backgrounds interact freely, making it easier to establish a credible cover story for how the recruiter and recruit came to know each other. Selecting a country restaurant or café with a good reputation—one that neither the recruiter nor the recruit has visited before—ensures that there is no previous association between the individuals and the location. Open verandas or secluded tables in garden areas provide additional discretion.

The most delicate part of the recruitment conversation should be conducted outside the restaurant, in a quiet outdoor setting such as a park or garden, following a shared meal or casual interaction. If necessary, the residency should organize surveillance of the meeting location, set up a signaling system between the recruiter and field operatives, and prepare transport for a quick exit should any signs of provocation or unexpected danger arise.

When selecting a recruitment location, intelligence officers must verify that the site is not regularly monitored by counterintelligence agencies or equipped with surveillance technology. The recruiter must also ensure that the location is not commonly used by capitalist intelligence services for meetings with their agents, as this could pose significant operational risks.

In certain cases, recruitment conversations may be conducted at the recruiter's apartment, provided that operational security requirements are met. This approach is more suitable when recruiting close personal connections or relatives who are already familiar with the recruiter's residence. However, the apartments of illegal intelligence officers or legal residency operatives should never be used for recruitment conversations. Given the need for an illegal intelligence officer to maintain absolute secrecy regarding their true identity, location, and profession, revealing their personal residence to a recruit—whose ultimate commitment is never guaranteed—would be a grave security breach.

For illegal operatives, conducting the recruitment conversation in a third country is often preferable. If both the recruiter and the recruit are foreigners in the location where the conversation takes place, the risk of adverse consequences is significantly reduced. If the recruit refuses to cooperate or reacts negatively, the intelligence officer has the ability to leave the country quickly, avoiding immediate danger or potential counterintelligence retaliation.

Developing a Recruitment Strategy

With the recruitment development process nearing its conclusion, several critical tasks must be finalized. The recruitment basis must be thoroughly refined, ensuring that the approach is sound and well-justified. The recruit's intelligence capabilities must be verified to determine their potential value and operational utility. Their personal qualities must be fully assessed to anticipate their likely reaction to the recruitment conversation. The selection of a recruiter must be carefully considered, ensuring that the individual conducting the recruitment is best suited to the task. Finally, the timing and location of the recruitment conversation must be strategically chosen to maximize the likelihood of success while minimizing security risks. .

Constructing a Comprehensive Recruitment Plan

When all necessary conditions for recruitment have been established—the recruitment basis has been selected and verified, the appropriate form and method have been determined, the recruit's personal qualities and intelligence capabilities have been thoroughly assessed, and a recruiter has been chosen—intelligence proceeds to the development of a formal recruitment plan. This plan serves as the blueprint for the final stage of recruitment development and includes all measures required to ensure a successful outcome.

The planning process begins with a comprehensive review of all intelligence gathered on the recruit. Intelligence officers must evaluate how thoroughly the candidate has been studied, confirming that their psychological profile, motivations, and vulnerabilities have been accurately assessed. This stage also involves verifying that the selected recruitment basis and method remain appropriate. The recruitment plan serves as a final analysis of all previous work, allowing intelligence officers to identify any weaknesses or errors in the development process and to take corrective measures before proceeding with the recruitment conversation. Various contingencies are considered, including potential complications during the conversation and strategies for overcoming resistance.

The recruitment plan briefly summarizes all relevant information about the candidate, detailing their intelligence capabilities and outlining the strategic value of their recruitment. The document provides a clear justification for the choice of recruitment basis and method, specifying any additional measures necessary to reinforce the selected approach. This justification must demonstrate how the intelligence service intends to ensure the successful completion of the recruitment process using the chosen form of engagement.

Another critical element of the plan is the rationale for selecting a particular recruiter. This section must demonstrate that the recruiter's personal characteristics, professional background, language skills, and access to the recruit have been carefully considered. The recruiter's ability to influence the recruit must be clearly established, ensuring that they are the best possible operative for the task. The plan also outlines the structure of the recruitment conversation, particularly in cases where a direct recruitment offer is being made. It must be evident from the plan that the recruiter will maintain a dominant position in the conversation, guiding its course while anticipating and countering any resistance from the recruit.

Finally, the plan includes contingency measures in case of a failed recruitment attempt. These measures are designed to protect the recruiter and any agents involved in the operation from exposure. The residency must anticipate various outcomes and establish procedures to prevent the failure from compromising broader intelligence activities. The recruitment plan, as a crucial operational document, is then sent to the central intelligence office, where it undergoes review and necessary modifications before final approval is granted.

In exceptional cases where immediate action is required, such as when there is no time to seek authorization from the central office or when communication with headquarters is disrupted, the resident may be granted the authority to sanction the recruitment interview independently. In such instances, the resident is required to submit a detailed report explaining the circumstances that necessitated the decision and providing a full account of the recruitment attempt and its outcome.

Executing and Formalizing the Recruitment Agreement

The culmination of recruitment development is the recruitment conversation, a unique and carefully orchestrated interaction. No two recruitment conversations are identical, just as no two recruits or recruiters are the same. Each recruitment conversation is shaped by the specific circumstances of the recruit, the operational environment, and the chosen recruitment approach.

When a recruit has been gradually drawn into intelligence work, the recruitment conversation serves as a formalization of their existing cooperation. It confirms their willingness to continue intelligence work under more secure conditions and with more advanced methods of communication. If the recruitment is based on ideological and political alignment, the conversation typically begins with a discussion of political topics, leading into the necessity of taking a more active role in the struggle. The recruiter then introduces the idea that covert work is the most effective way to achieve the recruit's goals and that cooperation with the intelligence service of a socialist state is aligned with their national interests. The conversation shifts towards how the recruit can best contribute, and finally, practical details such as methods of secret communication and meeting procedures are agreed upon. In many cases, the recruiter must also address the recruit's internal moral conflicts regarding secret cooperation, reframing intelligence work as a necessary and honorable extension of their existing beliefs.

When recruitment is conducted through a direct proposal, the conversation is typically more complex and presents a greater likelihood of resistance. A recruit approached with a direct offer may react in unpredictable ways, requiring the recruiter to be highly adaptable and prepared for a range of responses. Regardless of the recruitment method, it is essential to create a calm and controlled atmosphere before initiating the conversation. The recruiter must carefully observe the recruit's reactions, adjusting the approach as necessary to maintain the upper hand.

Throughout the conversation, the recruiter must exhibit tact, ensuring that the recruit is given time to express their thoughts and concerns. Argumentation must be presented convincingly but without diminishing the recruit's self-respect. Certain terms should be avoided to prevent negative connotations; for example, the recruit should be referred to as a "friend," "comrade," or "fighter for a common cause" rather than an "agent." Monetary compensation should be framed as financial assistance, reimbursement for expenses, or a token of appreciation rather than a direct reward. Similarly, cooperation with intelligence should be described as "helping our country" rather than engaging in espionage.

If the conversation takes place over a meal, care must be taken to accommodate the recruit's preferences while avoiding excessive alcohol consumption. While some intelligence officers believe that intoxication may make a recruit more susceptible to persuasion, this is a dangerous misconception. A recruit who agrees to cooperate while intoxicated may later reject their commitments, forget what was discussed, or worse, reveal the encounter to others in an unguarded moment. A recruitment conversation should always take place in a setting where both the recruiter and the recruit are fully conscious and aware of the significance of their discussion.

The recruiter must be well-prepared in advance, knowing exactly what promises can be made and what guarantees cannot be offered. Questions regarding financial compensation, potential relocation to a socialist country in case of danger, pension arrangements, or future benefits must be addressed with certainty. Any promises made must be realistic and enforceable, as failing to deliver on commitments can lead to the recruit's disillusionment and possible betrayal. The recruitment conversation should be dynamic, engaging, and, in most cases, conducted in a friendly manner. The exception to this rule applies to recruitments based on coercion or the use of compromising materials, where a different tone may be necessary.

During the conversation, the recruit must be assured that the risks associated with intelligence work are minimized through proper operational security and that the intelligence service has a vested interest in maintaining complete secrecy regarding their cooperation. In some cases, particularly when recruitment is based on coercion, the recruit may be asked to sign a written commitment to cooperate with the intelligence service. However, written agreements are generally avoided unless absolutely necessary, as an agent is bound to intelligence not by written commitments but by the secret information they provide. In such cases, financial receipts for payments received serve as an additional means of securing their cooperation, framed as standard financial documentation for the recruiter's accountability.

Once recruitment is completed, it is advisable to obtain a written autobiography and a photograph of the recruit. However, this should be done at an appropriate stage, preferably after the recruit has completed initial tasks and is more firmly integrated into intelligence work. A photograph is essential in case the recruit must be contacted by another intelligence officer, while a written autobiography provides critical background details on the recruit's family, social connections, and professional environment. The information contained in the autobiography allows intelligence officers to cross-check the recruit's statements against previous intelligence reports, ensuring that there are no discrepancies that might indicate deception.

The signing of receipts, written commitments, or other documentation provides an additional layer of security, reinforcing the recruit's commitment to continued cooperation. However, past experience in intelligence work has shown that counterintelligence operatives planted within intelligence networks readily sign agreements, provide written commitments, and give verbal assurances as a means of gaining the trust of intelligence officers. Because of this, a recruit cannot be considered fully committed to intelligence work simply based on their verbal or written agreement. A recruit's true loyalty and reliability are only confirmed once they have successfully completed a series of intelligence tasks, demonstrated their dependability, and developed a dependency on the intelligence service.

A recruitment conversation must always be conducted with a clear and purposeful objective. The recruiter must enter the conversation knowing exactly what task the recruit will be assigned upon completion of their recruitment. This is essential, as any recruit will inevitably ask why they were selected and what specific role the intelligence service expects them to fulfill. The nature of the first task assigned to the recruit is particularly important, as it must reinforce the idea that their cooperation is both necessary and valuable. The first task should be challenging enough to engage the recruit, yet achievable within their current capabilities. It must be concrete and clearly defined, ensuring that the recruit fully understands their role and the expectations placed upon them..

Ensuring Operational Security Throughout the Recruitment Process

In the process of recruitment development, maintaining strict secrecy and vigilance is of utmost importance. When assessing potential recruits, intelligence officers must obtain all necessary information without revealing their interest to the individuals involved in the evaluation process. Any uncalculated conversation or careless questioning may inadvertently expose the intelligence officer's intentions, leading to suspicion or even counterintelligence attention. Agents assisting in the study of a recruit should not become aware of the intelligence officer's true objective unless their role requires them to act as the recruiter. If it becomes necessary to disclose the nature of the recruitment effort to an agent, it should only be done with trusted individuals and under controlled circumstances.

All information gathered from agents and other sources must undergo rigorous verification and analysis, regardless of how credible the source appears to be. Intelligence officers must pay close attention to how an agent completes their task—whether they follow instructions precisely or deviate from assigned methods based on their own judgment. Even minor discrepancies or seemingly insignificant actions must be thoroughly reviewed. A careless or untrained agent may inadvertently draw the attention of counterintelligence, jeopardizing both themselves and the larger intelligence operation. An agent may also be recruited by enemy intelligence or, without realizing it, be manipulated by hostile counterintelligence services to expose the activities of socialist intelligence.

Despite thorough preparation, exhaustive study, and extensive recruitment development, the recruiter must always be prepared for the possibility that the recruit will refuse to cooperate. Even candidates who have already performed tasks for intelligence may not fully grasp the extent of their involvement and may resist formal recruitment. The risk associated with a failed recruitment attempt primarily lies in the fact that the recruit will now recognize the recruiter as an intelligence officer. This knowledge can lead to counterintelligence exposure, putting both the recruiter and any associated agents in danger if precautionary measures are not taken.

Recruits who have been drawn into intelligence work gradually on ideological-political or material grounds may still refuse formal cooperation due to personal fears, doubts, or concerns about failure. However, such individuals are less likely to immediately turn to the police or counterintelligence, as they understand that their prior contact with socialist intelligence could be used against them. The greater danger arises when individuals recruited via a direct proposal, particularly those motivated by material interests or those targeted with compromising materials, reject the offer. These recruits are often ideologically opposed to socialist intelligence and, despite their vulnerabilities, may decide to report the recruitment attempt to authorities, creating serious risks for the intelligence operation.

Experience in recruitment operations demonstrates that a well-prepared recruitment attempt and a properly structured recruitment conversation can often overcome a recruit's initial resistance. However, if it becomes evident during the conversation that the recruit's refusal is firm and applying additional pressure may push them toward taking hostile action, the recruiter must be prepared to withdraw. In such cases, the conversation should be redirected, and the recruiter must skillfully shift the discussion to convince the recruit that they misunderstood the nature of the proposal, thereby diffusing any immediate hostility.

An example of this strategic retreat occurred in an eastern country where a socialist intelligence officer had cultivated a friendship with a European diplomat known as "Ron." Ron had extensive connections among diplomats and frequently spoke positively about the Soviet Union and socialist countries. Background intelligence confirmed that Ron's sympathies were genuine. Over time, the intelligence officer strengthened his relationship with Ron, meeting regularly at diplomatic receptions where Ron would share confidential political insights under the guise of informal discussions among colleagues.

Convinced that Ron was a viable recruitment target, the intelligence officer obtained authorization to attempt recruitment. During a conversation, the officer suggested that, as a member of his diplomatic mission, he required a steady stream of political information but faced limitations due to his lower-ranking position and personal responsibilities. He proposed that Ron, as a trusted friend, assist by providing insights into political affairs. Ron, viewing the request as a casual favor, agreed to help.

A short time later, the intelligence officer suggested making their arrangement more discreet and hinted at financial compensation for Ron's contributions. Ron immediately took offense, stating that he found such a proposal insulting and that if this was about intelligence cooperation, he had no choice but to sever ties with the recruiter. Recognizing his mistake, the intelligence officer quickly pivoted. He insisted that Ron had misunderstood the request, emphasizing that the conversation had been about ensuring that Ron's assistance remained confidential for his own protection. He downplayed the mention of compensation, reframing it as a token of appreciation rather than a bribe. The sincerity of the officer's retreat, coupled with his convincing argument, calmed Ron. Though he likely remained somewhat suspicious, the situation was successfully de-escalated, allowing the intelligence officer to maintain their relationship and continue gathering political insights from Ron, albeit unofficially.

This case highlights two key points. First, even when a recruit has been gradually prepared for intelligence work, they may still refuse formal recruitment. Second, a refusal does not necessarily mean the operation has failed or that the recruit has become a security risk. The danger that arises from a recruit's refusal can often be neutralized through careful handling of the conversation, allowing the intelligence officer to preserve operational security.

Precautionary measures must be in place for every recruitment operation to address potential failures. A recruitment conversation should always be conducted in an environment that minimizes risks. The chosen location must ensure that the recruit cannot immediately report to the police or counterintelligence, summon witnesses, or create a public scandal. At the same time, the recruiter must have a clear exit strategy to avoid confrontation with local authorities or bystanders.

If a recruit refuses cooperation, the recruiter must skillfully retreat. They should work to convince the recruit that they misunderstood the proposal and attempt to de-escalate any tension that arises. If the recruit was targeted with compromising materials, the recruiter may suggest that the issue be forgotten— on the condition that the recruit maintains neutrality and does not act against the intelligence service. In extreme cases where there is a risk of denunciation, the recruit must be subtly reminded that they are dealing with a determined organization that will take protective measures if necessary.

In situations where a recruit's refusal appears to be part of a counterintelligence provocation, intelligence officers must immediately halt active operations, suspend meetings with agents, and focus on identifying potential security breaches. The recruit's actions and affiliations should be closely monitored to determine whether they have reported the recruitment attempt. Depending on the severity of the risk, it may be necessary for the recruiter to leave the country or be reassigned elsewhere to prevent further exposure.

The risk of recruitment failure cannot be entirely eliminated, but by implementing strategic countermeasures and maintaining flexibility, intelligence services can minimize the potential damage and, in many cases, preserve operational security even in the face of an unsuccessful recruitment attempt.

Integrating a Newly Recruited Agent into Intelligence Operations

Once a recruit has been successfully brought into

intelligence work, the recruiter—whether an intelligence officer or an agent—must immediately establish secure, confidential, and reliable communication with the new agent. Maintaining secrecy and control during this initial phase is critical, as the recruit is not yet fully integrated into the intelligence network and has not undergone extensive testing to confirm their reliability.

For intelligence officers operating from legal residencies, newly recruited agents typically remain in direct contact with the recruiter who brought them into the operation. The same applies to illegal intelligence officers if they were the ones to conduct the recruitment. This method ensures that, in the early stages of intelligence work, only one intelligence officer is compromised in the recruit's awareness as a true intelligence operative. Limiting exposure at this stage reduces the risk of counterintelligence penetration in case the recruit later proves unreliable.

In many cases, particularly when agents are recruited by experienced agent-recruiters rather than intelligence officers, the recruiter continues to act as the agent's primary contact and assumes the role of a group leader. This approach ensures continuity, as the recruit already has an established relationship with their recruiter, making the transition to operational intelligence work more seamless.

When intelligence officers from the central intelligence apparatus or recruiters from neighboring countries are involved in recruitment, newly recruited agents are often transferred to established intelligence officers or group leaders operating in the country where they will be conducting their work. This transition requires careful planning. The recruiter arranges a follow-up meeting with the new agent to confirm their commitment and ensure their understanding of the operational procedures. During this meeting, backup locations and alternative meeting points are agreed upon, contingency plans are discussed, and the recruit is given instructions on maintaining secrecy during travel to and from meeting locations. Every detail of the communication process is carefully structured to ensure that the agent remains under control while avoiding suspicion.

As the recruit begins intelligence work, these initial

communication procedures are further refined by the intelligence officer or group leader who assumes long-term responsibility for the recruit. This officer will supervise all aspects of the recruit's activities, ensuring that they develop into a valuable and reliable asset.

New agents who have not yet been fully assigned to specific intelligence tasks must be closely monitored. Intelligence officers working from both legal and illegal positions must remain highly vigilant regarding the behavior and performance of freshly recruited agents. Experience has shown that enemy intelligence services frequently attempt to infiltrate socialist intelligence networks by planting foreign intelligence agents as apparent recruits. These double agents readily agree to recruitment, make enthusiastic promises, and appear eager to establish trust. However, once embedded within the intelligence network, they attempt to introduce disinformation under the guise of valuable intelligence while subtly evading or delaying their assigned tasks. Their ultimate goal is to undermine socialist intelligence operations, mislead handlers, and gather information on operational methods, personnel, and objectives.

To prevent such infiltrations, intelligence officers must employ stringent verification measures, assessing each recruit's performance through carefully designed tasks that can confirm their legitimacy. New agents must be repeatedly tested for their commitment, reliability, and discretion. Only after proving their loyalty and effectiveness in real intelligence work can a recruit be fully integrated into the intelligence network. Recognizing and neutralizing enemy setups at an early stage is essential to maintaining the security and operational integrity of socialist intelligence services.

CONCLUSION

All intelligence operations conducted in capitalist countries, particularly those related to recruitment, require intelligence officers from socialist states to engage in the continuous and active search for suitable candidates and to secure their cooperation as agents. This task is not incidental but an essential function of socialist intelligence, aimed at strengthening its capacity to gather information and influence developments in adversarial states.

It must always be borne in mind that socialist states have access to a vast recruiting contingent within capitalist countries. The political, social, and economic conditions in these states provide intelligence officers with ample opportunities to identify, assess, and recruit individuals who possess the necessary intelligence capabilities. These conditions allow socialist intelligence services to systematically expand and consolidate a network of reliable and committed agents. The fulfillment of this objective is a direct requirement of the decision-making bodies of socialist governments, which depend on the steady acquisition of human assets to advance their strategic objectives.

At the same time, intelligence officers must recognize that their adversaries—enemy intelligence, counterintelligence, and police agencies—are constantly engaged in counterintelligence efforts designed to thwart their work. A primary objective of these agencies is to prevent socialist intelligence from successfully recruiting agents. To this end, capitalist counterintelligence operations seek to identify individuals under surveillance by socialist operatives and exploit any errors, lapses in discipline, or procedural missteps by intelligence officers or their agents. A single mistake, no matter how minor, can lead to the compromise of an entire intelligence network.

To achieve success in their work, intelligence officers must rigorously apply revolutionary Marxist-Leninist principles in practice, maintaining the highest levels of vigilance, discipline, and ideological commitment. Their ability to navigate complex and hostile environments depends on their understanding of dialectical and historical materialism as applied to intelligence work. Only through a combination of political conviction and technical expertise can they effectively counter the strategies of enemy intelligence agencies.

Intelligence officers must display initiative, adaptability, and constant dedication to refining their techniques and methods. The ever-evolving nature of intelligence work demands that they remain innovative in their approach, particularly in recruitment operations. The ability to develop and implement new strategies for identifying, assessing, and securing the cooperation of potential agents is crucial for overcoming the challenges posed by enemy counterintelligence.

Furthermore, intelligence officers must take full advantage of the political, economic, and social conditions in capitalist countries to facilitate recruitment efforts. By carefully analyzing the prevailing circumstances, they must develop and implement optimal strategies for identifying and engaging potential recruits, ensuring that recruitment operations are executed with the highest probability of success. This means finding ways to circumvent the barriers imposed by capitalist intelligence agencies, effectively navigating the operational landscape, and securing the most valuable agents within key institutions and facilities of capitalist states.

Agents embedded in strategic positions are the primary weapon of socialist intelligence in the struggle against the enemies of socialism and democracy. Their work is indispensable in gathering intelligence, shaping political developments, and countering the influence of imperialist forces. The success of socialist intelligence depends not only on the individual skills of its officers but on their ability to systematically build and maintain a vast, loyal, and effective agent network. In this way, intelligence services fulfill their historical role in advancing the socialist cause and defending the interests of the socialist community.

www.ingramcontent.com/pod-product-compliance
Lightning Source LLC
Chambersburg PA
CBHW062100270326
41931CB00013B/3151